THE WORLD AS I SEE IT

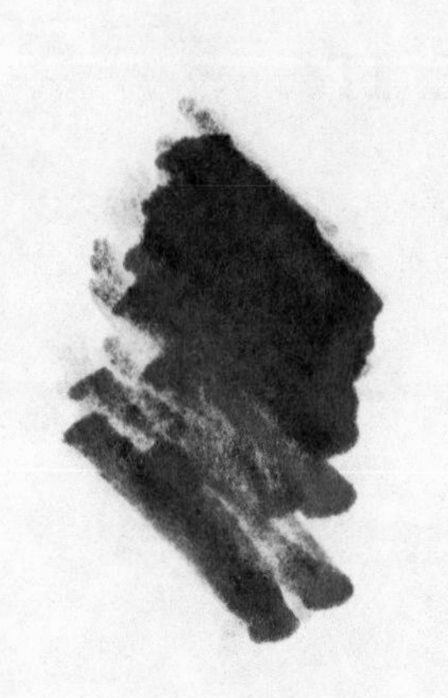

The World As I See It

ALBERT EINSTEIN

TRANSLATED BY
ALAN HARRIS

A PHILOSOPHICAL LIBRARY BOOK

This book is the authorized English translation of the volume 'Mein Weltbild' by Albert Einstein

Distributed to the trade by
CITADEL PRESS
A division of Lyle Stuart, Inc.
120 Enterprise Ave., Secaucus, N.J. 07094

PREFACE TO ORIGINAL EDITION

Only individuals have a sense of responsibility.—NIETZSCHE.

THIS book does not represent a complete collection of the articles, addresses, and pronouncements of Albert Einstein; it is a selection made with a definite object—namely, to give a picture of a man. To-day this man is being drawn, contrary to his own intention, into the whirlpool of political passions and contemporary history. As a result, Einstein is experiencing the fate that so many of the great men of history experienced: his character and opinions are being exhibited to the world in an utterly distorted form.

To forestall this fate is the real object of this book. It meets a wish that has constantly been expressed both by Einstein's friends and by the wider public. It contains work belonging to the most various dates—the article on "The International of Science" dates from the year 1922, the address on "The Principles of Scientific Research" from 1923, the "Letter to an Arab" from 1930—and the most various spheres, held together by the unity of the personality which stands behind all these utterances. Albert Einstein believes in humanity, in a peaceful world of mutual helpfulness, and in the high mission of science. This book is intended as a plea for this belief at a time which compels every one of us to overhaul his mental attitude and his ideas.

J. H.

INTRODUCTION TO ABRIDGED EDITION

In his biography of Einstein Mr. H. Gordon Garbedian relates that an American newspaper man asked the great physicist for a definition of his theory of relativity in one sentence. Einstein replied that it would take him three days to give a short definition of relativity. He might well have added that unless his questioner had an intimate acquaintance with mathematics and physics, the definition would be incomprehensible.

To the majority of people Einstein's theory is a complete mystery. Their attitude towards Einstein is like that of Mark Twain towards the writer of a work on mathematics: here was a man who had written an entire book of which Mark could not understand a single sentence. Einstein, therefore, is great in the public eye partly because he has made revolutionary discoveries which cannot be translated into the common tongue. We stand in proper awe of a man whose thoughts move on heights far beyond our range, whose achievements can be measured only by the few who are able to follow his reasoning and challenge his conclusions.

There is, however, another side to his personality. It is revealed in the addresses, letters, and occasional writings brought together in this book. These fragments form a mosaic portrait of Einstein the man. Each one is, in a sense, complete in itself; it presents his views on some aspect of progress, education, peace, war, liberty, or other problems of universal interest. Their combined effect is to demonstrate that the Einstein we can all understand is no less great than the Einstein we take on trust.

Einstein has asked nothing more from life than the freedom to pursue his researches into the mechanism of the universe. His nature is of rare simplicity and sincerity; he always has been, and he remains, genuinely indifferent to wealth and fame and the other prizes so dear to ambition. At the same time he is no recluse, shutting himself off from the sorrows and agitations of the world around him. Himself familiar from early years with the handicap of poverty and with some of the worst forms of man's inhumanity to man, he has never spared himself in defence of the weak and the oppressed. Nothing could be more unwelcome to his sensitive and retiring character than the glare of the platform and the heat of public controversy, yet he has never hesitated when he felt that his voice or influence would help to redress a wrong. History, surely, has few parallels with this introspective mathematical genius who laboured unceasingly as an eager champion of the rights of man.

Albert Einstein was born in 1879 at Ulm. When he was four years old his father, who owned an electrochemical works, moved to Munich, and two years later the boy went to school, experiencing a rigid, almost military, type of discipline and also the isolation of a shy and contemplative Jewish child among Roman Catholics—factors which made a deep and enduring impression. From the point of view of his teachers he was an unsatisfactory pupil, apparently incapable of progress in languages, history, geography, and other primary subjects. His interest in mathematics was roused, not by his instructors, but by a Jewish medical student, Max Talmey, who gave him a book on geometry, and so set him upon a course of enthusiastic study which made him, at the age of fourteen, a better mathematician than his masters. At this stage also he began the study of philosophy, reading and re-reading the words of Kant and other metaphysicians.

Business reverses led the elder Einstein to make a fresh start in Milan, thus introducing Albert to the joys of a freer, sunnier life than had been possible in Germany. Necessity, however, made this holiday a brief one, and after a few months of freedom the preparation for a career began. It opened with an effort, backed by a certificate of mathematical proficiency given by a teacher in the Gymnasium at Munich, to obtain admission to the Polytechnic Academy at Zurich. A year passed in the study of necessary subjects which he had neglected for mathematics, but once admitted, the young Einstein became absorbed in the pursuit of science and philosophy and made astonishing progress. After five distinguished years at the Polytechnic he hoped to step into the post of assistant professor, but found that the kindly words of the professors who had stimulated the hope did not materialize.

Then followed a weary search for work, two brief interludes of teaching, and a stable appointment as examiner at the Confederate Patent Office at Berne. Humdrum as the work was, it had the double advantage of providing a competence and of leaving his mind free for the mathematical speculations which were then taking shape in the theory of relativity. In 1905 his first monograph on the theory was published in a Swiss scientific journal, the *Annalen der Physik.* Zurich awoke to the fact that it possessed a genius in the form of a patent office clerk, promoted him to be a lecturer at the University and four years later—in 1909—installed him as Professor.

His next appointment was (in 1911) at the University of Prague, where he remained for eighteen months. Following a brief return to Zurich, he went, early in 1914, to Berlin as a professor in the Prussian Academy of Sciences and director of the Kaiser Wilhelm Institute for Theoretical Physics. The period of the Great War was a trying time for Einstein, who could not conceal his ardent pacifism, but he found what solace he could in his studies. Later events brought him into the open and into many parts of the world, as an exponent not only of pacifism but also of world-disarmament and the cause of Jewry. To a man of such views, as passionately held as they were by Einstein, Germany under the Nazis was patently impossible. In 1933 Einstein made his famous declaration : "As long as I have any choice, I will stay only in a country where political liberty, toleration, and equality of all citizens before the law are the rule." For a time he was a homeless exile; after offers had come to him from Spain and France and Britain, he settled in Princeton as Professor of Mathematical and Theoretical Physics, happy in his work, rejoicing in a free environment, but haunted always by the tragedy of war and oppression.

The World As I See It, in its original form, includes essays by Einstein on relativity and cognate subjects. For reasons indicated above, these have been omitted in the present edition; the object of this reprint is simply to reveal to the general reader the human side of one of the most dominating figures of our day.

CONTENTS

PART I

THE WORLD AS I SEE IT

PART II
POLITICS AND PACIFISM

PART III

GERMANY 1933

PART IV

THE JEWS

I

The World As I See It

The Meaning of Life

What is the meaning of human life, or of organic life altogether? To answer this question at all implies a religion. Is there any sense then, you ask, in putting it? I answer, the man who regards his own life and that of his fellow-creatures as meaningless is not merely unfortunate but almost disqualified for life.

The World as I see it

What an extraordinary situation is that of us mortals! Each of us is here for a brief sojourn; for what purpose he knows not, though he sometimes thinks he feels it. But from the point of view of daily life, without going deeper, we exist for our fellow-men—in the first place for those on whose smiles and welfare all our happiness depends, and next for all those unknown to us personally with whose destinies we are bound up by the tie of sympathy. A hundred times every day I remind myself that my inner and outer life depend on the labours of other men, living and dead, and that I must exert myself in order to give in the same measure as I have received and am still receiving. I am strongly drawn to the simple life and am often oppressed by the feeling that I am engrossing an unnecessary amount of the labour of my fellow-men. I regard class differences as contrary to justice and, in the last resort, based on force. I also consider that plain living is good for everybody, physically and mentally.

In human freedom in the philosophical sense I am definitely a disbeliever. Everybody acts not only under external compulsion but also in accordance with inner necessity. Schopenhauer's saying, that " a man can do as he will, but not will as he will," has been an inspiration to me since my youth up, and a continual consolation and unfailing well-spring of patience in the face of the hardships of life, my own and others'. This feeling mercifully mitigates the sense of responsibility which so easily becomes paralysing, and it prevents us from taking ourselves and other people too seriously; it conduces to a view of life in which humour, above all, has its due place.

To inquire after the meaning or object of one's own existence or of creation generally has always seemed to me absurd from an objective point of view. And yet everybody has certain ideals which determine the direction of his endeavours and his judgments. In this sense I have never looked upon ease and happiness as ends in themselves—such an ethical basis I call more proper for a herd of swine. The ideals which have lighted me on my way and time after time given me new courage to face life cheerfully, have been Truth, Goodness, and Beauty. Without the sense of fellowship with men of like mind, of preoccupation with the objective, the eternally unattainable in the field of art and scientific research, life would have seemed to me empty. The ordinary objects of human endeavour—property, outward success, luxury—have always seemed to me contemptible.

My passionate sense of social justice and social

responsibility has always contrasted oddly with my pronounced freedom from the need for direct contact with other human beings and human communities. I gang my own gait and have never belonged to my country, my home, my friends, or even my immediate family, with my whole heart; in the face of all these ties I have never lost an obstinate sense of detachment, of the need for solitude—a feeling which increases with the years. One is sharply conscious, yet without regret, of the limits to the possibility of mutual understanding and sympathy with one's fellow-creatures. Such a person no doubt loses something in the way of geniality and light-heartedness; on the other hand, he is largely independent of the opinions, habits, and judgments of his fellows and avoids the temptation to take his stand on such insecure foundations.

My political ideal is that of democracy. Let every man be respected as an individual and no man idolized. It is an irony of fate that I myself have been the recipient of excessive admiration and respect from my fellows through no fault, and no merit, of my own. The cause of this may well be the desire, unattainable for many, to understand the one or two ideas to which I have with my feeble powers attained through ceaseless struggle. I am quite aware that it is necessary for the success of any complex undertaking that one man should do the thinking and directing and in general bear the responsibility. But the led must not be compelled, they must be able to choose their leader. An autocratic system of coercion, in my opinion, soon degenerates. For force always attracts men of low morality, and I believe it to be an invariable rule that tyrants of genius are succeeded by scoundrels. For this reason I have always been passionately opposed to systems such as we see in Italy and Russia to-day. The thing that has brought discredit upon the prevailing form of democracy in Europe to-day is not to be laid to the door of the democratic idea as such, but to lack of stability on the part of the heads of governments and to the impersonal character of the electoral system. I believe that in this respect the United States of America have found the right way. They have a responsible President who is elected for a sufficiently long period and has sufficient powers to be really responsible. On the other hand, what I value in our political system is the more extensive provision that it makes for the individual in case of illness or need. The really valuable thing in the pageant of human life seems to me not the State but the creative, sentient individual, the personality; it alone creates the noble and the sublime, while the herd as such remains dull in thought and dull in feeling.

This topic brings me to that worst outcrop of the herd nature, the military system, which I abhor. That a man can take pleasure in marching in formation to the strains of a band is enough to make me despise him. He has only been given his big brain by mistake; a backbone was all he needed. This plague-spot of civilization ought to be abolished with all possible speed. Heroism by order, senseless violence, and all the pestilent nonsense that goes by the name of patriotism—how I hate them! War seems to me a mean, contemptible thing: I would rather be hacked in pieces than take part in such an abominable business. And yet so high, in spite of everything, is my opinion of the human race that I believe this bogey would have disappeared long ago, had the sound sense of the nations not been systematically corrupted by commercial and political interests acting through the schools and the Press.

The fairest thing we can experience is the mysterious. It is the fundamental emotion which stands at the cradle of true art and true science. He who knows it not and can no longer wonder, no longer feel amazement, is as good as dead, a snuffed-out candle. It was the experience of mystery—even if mixed with fear—that engendered religion. A knowledge of the existence of something we cannot penetrate, of the manifestations of the profoundest reason and the most radiant beauty, which are only accessible to our reason in their most elementary forms—it is this knowledge and this emotion that constitute the truly religious attitude; in this sense, and in this alone, I am a deeply religious man. I cannot conceive of a God who rewards and punishes his creatures, or has a will of the type of which we are conscious in ourselves. An individual who should survive his physical death is also beyond my comprehension, nor do I wish it otherwise; such notions are for the fears or absurd egoism of feeble souls. Enough for me the mystery of the eternity of life, and the inkling of the marvellous structure of reality, together with the single-hearted endeavour to comprehend a portion, be it never so tiny, of the reason that manifests itself in nature.

The Liberty of Doctrine—à propos of the Gumbel Case

Academic chairs are many, but wise and noble teachers are few; lecture-rooms are numerous and large, but the number of young people who genuinely thirst after truth and justice is small.

Nature scatters her common wares with a lavish hand, but the choice sort she produces but seldom.

We all know that, so why complain? Was it not ever thus and will it not ever thus remain? Certainly, and one must take what Nature gives as one finds it. But there is also such a thing as a spirit of the times, an attitude of mind characteristic of a particular generation, which is passed on from individual to individual and gives a society its particular tone. Each of us has to do his little bit towards transforming this spirit of the times.

Compare the spirit which animated the youth in our universities a hundred years ago with that prevailing to-day. They had faith in the amelioration of human society, respect for every honest opinion, the tolerance for which our classics had lived and fought. In those days men strove for a larger political unity, which at that time was called Germany. It was the students and the teachers at the universities who kept these ideals alive.

To-day also there is an urge towards social progress, towards tolerance and freedom of thought, towards a larger political unity, which we to-day call Europe. But the students at our universities have ceased as completely as their teachers to enshrine the hopes and ideals of the nation. Anyone who looks at our times coolly and dispassionately must admit this.

We are assembled to-day to take stock of ourselves. The external reason for this meeting is the Gumbel case. This apostle of justice has written about unexpiated political crimes with devoted industry, high courage, and exemplary fairness, and has done the community a signal service by his books. And this is the man whom the students, and a good many of the staff, of his university are to-day doing their best to expel.

Political passion cannot be allowed to go to such lengths. I am convinced that every man who reads Herr Gumbel's books with an open mind will get the same impression from them as I have. Men like him are needed if we are ever to build up a healthy political society.

Let every man judge according to his own standards, by what he has himself read, not by what others tell him.

If that happens, this Gumbel case, after an unedifying beginning, may still do good.

Good and Evil

It is right in principle that those should be the best loved who have contributed most to the elevation of the human race and human life. But, if one goes on to ask who they are, one finds oneself in no inconsiderable difficulties. In the case of political, and even of religious, leaders, it is often very doubtful whether they have done more good or harm. Hence I most seriously believe that one does people the best service by giving them some elevating work to do and thus indirectly elevating them. This applies most of all to the great artist, but also in a lesser degree to the scientist. To be sure, it is not the fruits of scientific research that elevate a man and enrich his nature, but the urge to understand, the intellectual work, creative or receptive. It would surely be absurd to judge the value of the Talmud, for instance, by its intellectual fruits.

The true value of a human being is determined primarily by the measure and the sense in which he has attained to liberation from the self.

Society and Personality

When we survey our lives and endeavours we soon observe that almost the whole of our actions and desires are bound up with the existence of other human beings. We see that our whole nature resembles that of the social animals. We eat food that others have grown, wear clothes that others have made, live in houses that others have built. The greater part of our knowledge and beliefs has been communicated to us by other people through the medium of a language which others have created. Without language our mental capacities would be poor indeed, comparable to those of the higher animals; we have, therefore, to admit that we owe our principal advantage over the beasts to the fact of living in human society. The individual, if left alone from birth, would remain primitive and beast-like in his thoughts and feelings to a degree that we can hardly conceive. The individual is what he is and has the significance that he has not so much in virtue of his individuality, but rather as a member of a great human society, which directs his material and spiritual existence from the cradle to the grave.

A man's value to the community depends primarily on how far his feelings, thoughts, and actions are directed towards promoting the good of his fellows. We call him good or bad according to how he stands in this matter. It looks at first sight as if our estimate of a man depended entirely on his social qualities.

And yet such an attitude would be wrong. It

is clear that all the valuable things, material, spiritual, and moral, which we receive from society can be traced back through countless generations to certain creative individuals. The use of fire, the cultivation of edible plants, the steam engine—each was discovered by one man.

Only the individual can think, and thereby create new values for society—nay, even set up new moral standards to which the life of the community conforms. Without creative, independently thinking and judging personalities the upward development of society is as unthinkable as the development of the individual personality without the nourishing soil of the community.

The health of society thus depends quite as much on the independence of the individuals composing it as on their close political cohesion. It has been said very justly that Græco-Europeo-American culture as a whole, and in particular its brilliant flowering in the Italian Renaissance, which put an end to the stagnation of mediæval Europe, is based on the liberation and comparative isolation of the individual.

Let us now consider the times in which we live. How does society fare, how the individual? The population of the civilized countries is extremely dense as compared with former times; Europe to-day contains about three times as many people as it did a hundred years ago. But the number of great men has decreased out of all proportion. Only a few individuals are known to the masses as personalities, through their creative achievements. Organization has to some extent taken the place of the great man, particularly in the technical sphere, but also to a very perceptible extent in the scientific.

The lack of outstanding figures is particularly striking in the domain of art. Painting and music have definitely degenerated and largely lost their popular appeal. In politics not only are leaders lacking, but the independence of spirit and the sense of justice of the citizen have to a great extent declined. The democratic, parliamentarian regime, which is based on such independence, has in many places been shaken, dictatorships have sprung up and are tolerated, because men's sense of the dignity and the rights of the individual is no longer strong enough. In two weeks the sheep-like masses can be worked up by the newspapers into such a state of excited fury that the men are prepared to put on uniform and kill and be killed, for the sake of the worthless aims of a few interested parties. Compulsory military service seems to me the most disgraceful symptom of that deficiency in personal dignity from which civilized mankind is suffering to-day. No wonder there is no lack of prophets who prophesy the early eclipse of our civilization. I am not one of these pessimists; I believe that better times are coming. Let me shortly state my reasons for such confidence.

In my opinion, the present symptoms of decadence are explained by the fact that the development of industry and machinery has made the struggle for existence very much more severe, greatly to the detriment of the free development of the individual. But the development of machinery means that less and less work is needed from the individual for the satisfaction of the community's needs. A planned division of labour is becoming more and more of a crying necessity, and this division will lead to the material security of the individual. This security and the spare time and energy which the individual will have at his command can be made to further his development. In this way the community may regain its health, and we will hope that future historians will explain the morbid symptoms of present-day society as the childhood ailments of an aspiring humanity, due entirely to the excessive speed at which civilization was advancing.

Address at the Grave of H. A. Lorentz

It is as the representative of the German-speaking academic world, and in particular the Prussian Academy of Sciences, but above all as a pupil and affectionate admirer that I stand at the grave of the greatest and noblest man of our times. His genius was the torch which lighted the way from the teachings of Clerk Maxwell to the achievements of contemporary physics, to the fabric of which he contributed valuable materials and methods.

His life was ordered like a work of art down to the smallest detail. His never-failing kindness and magnanimity and his sense of justice, coupled with an intuitive understanding of people and things, made him a leader in any sphere he entered. Everyone followed him gladly, for they felt that he never set out to dominate but always simply to be of use. His work and his example will live on as an inspiration and guide to future generations.

H. A. Lorentz's work in the cause of International Co-operation

With the extensive specialization of scientific research which the nineteenth century brought about, it has become rare for a man occupying a leading position in one of the sciences to manage at the same time to do valuable service to the community in the sphere of international organization and international politics. Such service demands not only energy, insight, and a reputation based on solid achievements, but also a freedom from national prejudice and a devotion to the common ends of all, which have become rare in our times. I have met no one who combined all these qualities in himself so perfectly as H. A. Lorentz. The marvellous thing about the effect of his personality was this: Independent and headstrong natures, such as are particularly common among men of learning, do not readily bow to another's will and for the most part only accept his leadership grudgingly. But, when Lorentz is in the presidential chair, an atmosphere of happy co-operation is invariably created, however much those present may differ in their aims and habits of thought. The secret of this success lies not only in his swift comprehension of people and things and his marvellous command of language, but above all in this, that one feels that his whole heart is in the business in hand, and that, when he is at work, he has room for nothing else in his mind. Nothing disarms the recalcitrant so much as this.

Before the war Lorentz's activities in the cause of international relations were confined to presiding at congresses of physicists. Particularly noteworthy among these were the Solvay Congresses, the first two of which were held at Brussels in 1909 and 1912. Then came the European war, which was a crushing blow to all who had the improvement of human relations in general at heart. Even before the war was over, and still more after its end, Lorentz devoted himself to the work of reconciliation. His efforts were especially directed towards the re-establishment of fruitful and friendly co-operation between men of learning and scientific societies. An outsider can hardly conceive what uphill work this is. The accumulated resentment of the war period has not yet died down, and many influential men persist in the irreconcilable attitude into which they allowed themselves to be driven by the pressure of circumstances. Hence Lorentz's efforts resemble those of a doctor with a recalcitrant patient who refuses to take the medicines carefully prepared for his benefit.

But Lorentz is not to be deterred, once he has recognized a course of action as the right one. The moment the war was over, he joined the governing body of the "Conseil de recherche," which was founded by the savants of the victorious countries, and from which the savants and learned societies of the Central Powers were excluded. His object in taking this step, which caused great offence to the academic world of the Central Powers, was to influence this institution in such a way that it could be expanded into something truly international. He and other right-minded men succeeded, after repeated efforts, in securing the removal of the offensive exclusion-clause from the statutes of the "Conseil." The goal, which is the restoration of normal and fruitful co-operation between learned societies, is, however, not yet attained, because the academic world of the Central Powers, exasperated by nearly ten years of exclusion from practically all international gatherings, has got into a habit of keeping itself to itself. Now, however, there are good grounds for hoping that the ice will soon be broken, thanks to the tactful efforts of Lorentz, prompted by pure enthusiasm for the good cause.

Lorentz has also devoted his energies to the service of international cultural ends in another way, by consenting to serve on the League of Nations Commission for international intellectual co-operation, which was called into existence some five years ago with Bergson as chairman. For the last year Lorentz has presided over the Commission, which, with the active support of its subordinate, the Paris Institute, is to act as a go-between in the domain of intellectual and artistic work among the various spheres of culture. There too the beneficent influence of this intelligent, humane, and modest personality, whose unspoken but faithfully followed advice is, "Not mastery but service," will lead people in the right way.

May his example contribute to the triumph of that spirit!

In Honour of Arnold Berliner's Seventieth Birthday

(Arnold Berliner is the editor of the periodical *Die Naturwissenschaften.*)

I should like to take this opportunity of telling my friend Berliner and the readers of this paper why I rate him and his work so highly. It has to be done here because it is one's only chance of getting such things said; since our training in objectivity has led to a taboo on everything personal, which we mortals may transgress only on quite exceptional occasions such as the present one.

And now, after this dash for liberty, back to the objective! The province of scientifically determined fact has been enormously extended, theoretical knowledge has become vastly more profound in every department of science. But the assimilative power of the human intellect is and remains strictly limited. Hence it was inevitable that the activity of the individual investigator should be confined to a smaller and smaller section of human knowledge. Worse still, as a result of this specialization, it is becoming increasingly difficult for even a rough general grasp of science as a whole, without which the true spirit of research is inevitably handicapped, to keep pace with progress. A situation is developing similar to the one symbolically represented in the Bible by the story of the Tower of Babel. Every serious scientific worker is painfully conscious of this involuntary relegation to an ever-narrowing sphere of knowledge, which is threatening to deprive the investigator of his broad horizon and degrade him to the level of a mechanic.

We have all suffered under this evil, without making any effort to mitigate it. But Berliner has come to the rescue, as far as the German-speaking world is concerned, in the most admirable way. He saw that the existing popular periodicals were sufficient to instruct and stimulate the layman; but he also saw that a first-class, well-edited organ was needed for the guidance of the scientific worker who desired to be put sufficiently *au courant* of developments in scientific problems, methods, and results to be able to form a judgment of his own. Through many years of hard work he has devoted himself to this object with great intelligence and no less great determination, and done us all, and science, a service for which we cannot be too grateful.

It was necessary for him to secure the co-operation of successful scientific writers and induce them to say what they had to say in a form as far as possible intelligible to non-specialists. He has often told me of the fights he had in pursuing this object, the difficulties of which he once described to me in the following riddle: Question: What is a scientific author? Answer: A cross between a mimosa and a porcupine.[1] Berliner's achievement would have been impossible but for the peculiar intensity of his longing for a clear, comprehensive view of the largest possible area of scientific country. This feeling also drove him to produce a text-book of physics, the fruit of many years of strenuous work, of which a medical student said to me the other day: "I don't know how I should ever have got a clear idea of the principles of modern physics in the time at my disposal without this book."

Berliner's fight for clarity and comprehensiveness of outlook has done a great deal to bring the problems, methods, and results of science home to many people's minds. The scientific life of our time is simply inconceivable without his paper. It is just as important to make knowledge live and to keep it alive as to solve specific problems. We are all conscious of what we owe to Arnold Berliner.

Popper-Lynkæus was more than a brilliant engineer and writer. He was one of the few out-

[1] Do not be angry with me for this indiscretion, my dear Berliner. A serious-minded man enjoys a good laugh now and then.

standing personalities who embody the conscience of a generation. He has drummed it into us that society is responsible for the fate of every individual and shown us a way to translate the consequent obligation of the community into fact. The community or State was no fetish to him; he based its right to demand sacrifices of the individual entirely on its duty to give the individual personality a chance of harmonious development.

Obituary of the Surgeon, M. Katzenstein

During the eighteen years I spent in Berlin I had few close friends, and the closest was Professor Katzenstein. For more than ten years I spent my leisure hours during the summer months with him, mostly on his delightful yacht. There we confided our experiences, ambitions, emotions to each other. We both felt that this friendship was not only a blessing because each understood the other, was enriched by him, and found in him that responsive echo so essential to anybody who is truly alive; it also helped to make both of us more independent of external experience, to objectivize it more easily.

I was a free man, bound neither by many duties nor by harassing responsibilities; my friend, on the contrary, was never free from the grip of urgent duties and anxious fears for the fate of those in peril. If, as was invariably the case, he had performed some dangerous operations in the morning, he would ring up on the telephone, immediately before we got into the boat, to enquire after the condition of the patients about whom he was worried; I could see how deeply concerned he was for the lives entrusted to his care. It was marvellous that this shackled outward existence did not clip the wings of his soul; his imagination and his sense of humour were irrepressible. He never became the typical conscientious North German, whom the Italians in the days of their freedom used to call *bestia seriosa*. He was sensitive as a youth to the tonic beauty of the lakes and woods of Brandenburg, and as he sailed the boat with an expert hand through these beloved and familiar surroundings he opened the secret treasure-chamber of his heart to me—he spoke of his experiments, scientific ideas, and ambitions. How he found time and energy for them was always a mystery to me; but the passion for scientific enquiry is not to be crushed by any burdens. The man who is possessed with it perishes sooner than it does.

There were two types of problems that engaged his attention. The first forced itself on him out of the necessities of his practice. Thus he was always thinking out new ways of inducing healthy muscles to take the place of lost ones, by ingenious transplantation of tendons. He found this remarkably easy, as he possessed an uncommonly strong spatial imagination and a remarkably sure feeling for mechanism. How happy he was when he had succeeded in making somebody fit for normal life by putting right the muscular system of his face, foot, or arm! And the same when he avoided an operation, even in cases which had been sent to him by physicians for surgical treatment (in cases of gastric ulcer by neutralizing the pepsin). He also set great store by the treatment of peritonitis by an anti-toxic coli-serum which he discovered, and rejoiced in the successes he achieved with it. In talking of it he often lamented the fact that this method of treatment was not endorsed by his colleagues.

The second group of problems had to do with the common conception of an antagonism between different sorts of tissue. He believed that he was here on the track of a general biological principle of widest application, whose implications he followed out with admirable boldness and persistence. Starting out from this basic notion he discovered that osteomyelon and periosteum prevent each other's growth if they are not separated from each other by bone. In this way he succeeded in explaining hitherto inexplicable cases of wounds failing to heal, and in bringing about a cure.

This general notion of the antagonism of the tissues, especially of epithelium and connective tissue, was the subject to which he devoted his scientific energies, especially in the last ten years of his life. Experiments on animals and a systematic investigation of the growth of tissues in a nutrient fluid were carried out side by side. How thankful he was, with his hands tied as they were by his duties, to have found such an admirable and infinitely enthusiastic fellow-worker in Fräulein Knake! He succeeded in securing wonderful results bearing on the factors which favour the growth of epithelium at the expense of that of connective tissue, results which may well be of decisive importance for the study of cancer. He also had the pleasure of inspiring his own son to become his intelligent and independent fellow-worker, and of exciting the warm interest and co-operation of Sauerbruch just in the last years of his life, so that he was able to die with the consoling thought that his life's work would not perish, but would be vigorously continued on the lines he had laid down.

I for my part am grateful to fate for having given me this man, with his inexhaustible goodness and high creative gifts, for a friend.

Congratulations to Dr. Solf

I am delighted to be able to offer you, Dr. Solf, the heartiest congratulations, the congratulations of Lessing College, of which you have become an indispensable pillar, and the congratulations of all who are convinced of the need for close contact between science and art and the public which is hungry for spiritual nourishment.

You have not hesitated to apply your energies to a field where there are no laurels to be won, but quiet, loyal work to be done in the interests of the general standard of intellectual and spiritual life, which is in peculiar danger to-day owing to a variety of circumstances. Exaggerated respect for athletics, an excess of coarse impressions which the complications of life through the technical discoveries of recent years has brought with it, the increased severity of the struggle for existence due to the economic crisis, the brutalization of political life—all these factors are hostile to the ripening of the character and the desire for real culture, and stamp our age as barbarous, materialistic, and superficial. Specialization in every sphere of intellectual work is producing an ever-widening gulf between the intellectual worker and the non-specialist, which makes it more difficult for the life of the nation to be fertilized and enriched by the achievements of art and science.

But contact between the intellectual and the masses must not be lost. It is necessary for the elevation of society and no less so for renewing the strength of the intellectual worker; for the flower of science does not grow in the desert. For this reason you, Herr Solf, have devoted a portion of your energies to Lessing College, and we are grateful to you for doing so. And we wish you further success and happiness in your work for this noble cause.

Of Wealth

I am absolutely convinced that no wealth in the world can help humanity forward, even in the hands of the most devoted worker in this cause. The example of great and pure characters is the only thing that can produce fine ideas and noble deeds. Money only appeals to selfishness and always tempts its owners irresistibly to abuse it.

Can anyone imagine Moses, Jesus, or Gandhi armed with the money-bags of Carnegie?

Education and Educators

A letter.

Dear Miss——,

I have read about sixteen pages of your manuscript and it made me—smile. It is clever, well observed, honest, it stands on its own feet up to a point, and yet it is so typically feminine, by which I mean derivative and vitiated by personal rancour. I suffered exactly the same treatment at the hands of my teachers, who disliked me for my independence and passed me over when they wanted assistants (I must admit that I was somewhat less of a model student than you). But it would not have been worth my while to write anything about my school life, still less would I have liked to be responsible for anyone's printing or actually reading it. Besides, one always cuts a poor figure if one complains about others who are struggling for their place in the sun too after their own fashion.

Therefore pocket your temperament and keep your manuscript for your sons and daughters, in order that they may derive consolation from it and—not give a damn for what their teachers tell them or think of them.

Incidentally I am only coming to Princeton to research, not to teach. There is too much education altogether, especially in American schools. The only rational way of educating is to be an example—of what to avoid, if one can't be the other sort.

With best wishes.

To the Schoolchildren of Japan

In sending this greeting to you Japanese schoolchildren, I can lay claim to a special right to do so. For I have myself visited your beautiful country, seen its cities and houses, its mountains and woods, and in them Japanese boys who had learnt from them to love their country. A big fat book full of coloured drawings by Japanese children lies always on my table.

If you get my message of greeting from all this distance, bethink you that ours is the first age in history to bring about friendly and understanding intercourse between people of different countries; in former times nations passed their lives in mutual ignorance, and in fact hated or feared one another. May the spirit of brotherly understanding gain ground more and more among them. With this in mind I, an old man, greet you Japanese schoolchildren from afar and hope that your generation may some day put mine to shame.

Teachers and Pupils

An address to children

(The principal art of the teacher is to awaken the joy in creation and knowledge.)

My dear Children,

I rejoice to see you before me to-day, happy youth of a sunny and fortunate land.

Bear in mind that the wonderful things you learn in your schools are the work of many generations, produced by enthusiastic effort and infinite labour in every country of the world. All this is put into your hands as your inheritance in order that you may receive it, honour it, add to it, and one day faithfully hand it on to your children. Thus do we mortals achieve immortality in the permanent things which we create in common.

If you always keep that in mind you will find a meaning in life and work and acquire the right attitude towards other nations and ages.

Paradise Lost

As late as the seventeenth century the savants and artists of all Europe were so closely united by the bond of a common ideal that co-operation between them was scarcely affected by political events. This unity was further strengthened by the general use of the Latin language.

To-day we look back at this state of affairs as at a lost paradise. The passions of nationalism have destroyed this community of the intellect, and the Latin language, which once united the whole world, is dead. The men of learning have become the chief mouthpieces of national tradition and lost their sense of an intellectual commonwealth.

Nowadays we are faced with the curious fact that the politicians, the practical men of affairs, have become the exponents of international ideas. It is they who have created the League of Nations.

Religion and Science

Everything that the human race has done and thought is concerned with the satisfaction of felt needs and the assuagement of pain. One has to keep this constantly in mind if one wishes to understand spiritual movements and their development. Feeling and desire are the motive forces behind all human endeavour and human creation, in however exalted a guise the latter may present itself to us. Now what are the feelings and needs that have led men to religious thought and belief in the widest sense of the words? A little consideration will suffice to show us that the most varying emotions preside over the birth of religious thought and experience. With primitive man it is above all fear that evokes religious notions—fear of hunger, wild beasts, sickness, death. Since at this stage of existence understanding of causal connexions is usually poorly developed, the human mind creates for itself more or less analogous beings on whose wills and actions these fearful happenings depend. One's object now is to secure the favour of these beings by carrying out actions and offering sacrifices which, according to the tradition handed down from generation to generation, propitiate them or make them well disposed towards a mortal. I am speaking now of the religion of fear. This, though not created, is in an important degree stabilized by the formation of a special priestly caste which sets up as a mediator between the people and the beings they fear, and erects a hegemony on this basis. In many cases the leader or ruler whose position depends on other factors, or a privileged class, combines priestly functions with its secular authority in order to make the latter more secure; or the political rulers and the priestly caste make common cause in their own interests.

The social feelings are another source of the crystallization of religion. Fathers and mothers and the leaders of larger human communities are mortal and fallible. The desire for guidance, love, and support prompts men to form the social or moral conception of God. This is the God of Providence who protects, disposes, rewards, and punishes, the God who, according to the width of the believer's outlook, loves and cherishes the life of the tribe or of the human race, or even life as such, the comforter in sorrow and unsatisfied longing, who preserves the souls of the dead. This is the social or moral conception of God.

The Jewish scriptures admirably illustrate the development from the religion of fear to moral religion, which is continued in the New Testament. The religions of all civilized peoples, especially the peoples of the Orient, are primarily moral religions. The development from a religion of fear to moral religion is a great step in a nation's life. That primitive religions are based entirely on fear and the religions of civilized peoples purely on morality is a prejudice against which we must be on our guard. The truth is that they are all intermediate types, with this reservation, that on the higher levels of social life the religion of morality predominates.

Common to all these types is the anthropomorphic character of their conception of God. Only individuals of exceptional endowments and exceptionally high-minded communities, as a general rule, get in any real sense beyond this level. But there is a third state of religious experience which belongs to all of them, even though it is rarely found in a pure form, and which I will call cosmic religious feeling. It is very difficult to explain this feeling to anyone who is entirely without it, especially as there is no anthropomorphic conception of God corresponding to it.

The individual feels the nothingness of human desires and aims and the sublimity and marvellous order which reveal themselves both in nature and in the world of thought. He looks upon individual existence as a sort of prison and wants to experience the universe as a single significant whole. The beginnings of cosmic religious feeling already appear in earlier stages of development—e.g., in many of the Psalms of David and in some of the Prophets. Buddhism, as we have learnt from the wonderful writings of Schopenhauer especially, contains a much stronger element of it.

The religious geniuses of all ages have been distinguished by this kind of religious feeling, which knows no dogma and no God conceived in man's image; so that there can be no Church whose central teachings are based on it. Hence it is precisely among the heretics of every age that we find men who were filled with the highest kind of religious feeling and were in many cases regarded by their contemporaries as Atheists, sometimes also as saints. Looked at in this light, men like Democritus, Francis of Assisi, and Spinoza are closely akin to one another.

How can cosmic religious feeling be communicated from one person to another, if it can give rise to no definite notion of a God and no theology? In my view, it is the most important function of art and science to awaken this feeling and keep it alive in those who are capable of it.

We thus arrive at a conception of the relation of science to religion very different from the usual one. When one views the matter historically one is inclined to look upon science and religion as irreconcilable antagonists, and for a very obvious reason. The man who is thoroughly convinced of the universal operation of the law of causation cannot for a moment entertain the idea of a being who interferes in the course of events—that is, if he takes the hypothesis of causality really seriously. He has no use for the religion of fear and equally little for social or moral religion. A God who rewards and punishes is inconceivable to him for the simple reason that a man's actions are determined by necessity, external and internal, so that in God's eyes he cannot be responsible, any more than an inanimate object is responsible for the motions it goes through. Hence science has been charged with undermining morality, but the charge is unjust. A man's ethical behaviour should be based effectually on sympathy, education, and social ties; no religious basis is necessary. Man would indeed be in a poor way if he had to be restrained by fear and punishment and hope of reward after death.

It is therefore easy to see why the Churches have

always fought science and persecuted its devotees. On the other hand, I maintain that cosmic religious feeling is the strongest and noblest incitement to scientific research. Only those who realize the immense efforts and, above all, the devotion which pioneer work in theoretical science demands, can grasp the strength of the emotion out of which alone such work, remote as it is from the immediate realities of life, can issue. What a deep conviction of the rationality of the universe and what a yearning to understand, were it but a feeble reflection of the mind revealed in this world, Kepler and Newton must have had to enable them to spend years of solitary labour in disentangling the principles of celestial mechanics! Those whose acquaintance with scientific research is derived chiefly from its practical results easily develop a completely false notion of the mentality of the men who, surrounded by a sceptical world, have shown the way to those like-minded with themselves, scattered through the earth and the centuries. Only one who has devoted his life to similar ends can have a vivid realization of what has inspired these men and given them the strength to remain true to their purpose in spite of countless failures. It is cosmic religious feeling that gives a man strength of this sort. A contemporary has said, not unjustly, that in this materialistic age of ours the serious scientific workers are the only profoundly religious people.

The Religiousness of Science

You will hardly find one among the profounder sort of scientific minds without a peculiar religious feeling of his own. But it is different from the religion of the naive man. For the latter God is a being from whose care one hopes to benefit and whose punishment one fears; a sublimation of a feeling similar to that of a child for its father, a being to whom one stands to some extent in a personal relation, however deeply it may be tinged with awe.

But the scientist is possessed by the sense of universal causation. The future, to him, is every whit as necessary and determined as the past. There is nothing divine about morality, it is a purely human affair. His religious feeling takes the form of a rapturous amazement at the harmony of natural law, which reveals an intelligence of such superiority that, compared with it, all the systematic thinking and acting of human beings is an utterly insignificant reflection. This feeling is the guiding principle of his life and work, in so far as he succeeds in keeping himself from the shackles of selfish desire. It is beyond question closely akin to that which has possessed the religious geniuses of all ages.

The Plight of Science

The German-speaking countries are menaced by a danger to which those in the know are in duty bound to call attention in the most emphatic terms. The economic stress which political events bring in their train does not hit everybody equally hard. Among the hardest hit are the institutions and individuals whose material existence depends directly on the State. To this category belong the scientific institutions and workers on whose work not merely the well-being of science but also the position occupied by Germany and Austria in the scale of culture very largely depends.

To grasp the full gravity of the situation it is necessary to bear in mind the following consideration. In times of crisis people are generally blind to everything outside their immediate necessities. For work which is directly productive of material wealth they will pay. But science, if it is to flourish, must have no practical end in view. As a general rule, the knowledge and the methods which it creates only subserve practical ends indirectly and, in many cases, not till after the lapse of several generations. Neglect of science leads to a subsequent dearth of intellectual workers able, in virtue of their independent outlook and judgment, to blaze new trails for industry or adapt themselves to new situations. Where scientific enquiry is stunted the intellectual life of the nation dries up, which means the withering of many possibilities of future development. This is what we have to prevent. Now that the State has been weakened as a result of non-political causes, it is up to the economically stronger members of the community to come to the rescue directly, and prevent the decay of scientific life.

Far-sighted men with a clear understanding of the situation have set up institutions by which scientific work of every sort is to be kept going in Germany and Austria. Help to make these efforts a real success. In my teaching work I see with admiration that economic troubles have not yet succeeded in stifling the will and the enthusiasm for scientific research. Far from it! Indeed, it looks as if our disasters had actually quickened the devotion to non-material goods. Everywhere people are working with burning enthusiasm in the most difficult circumstances.

See to it that the will-power and the talents of the youth of to-day do not perish to the grievous hurt of the community as a whole.

Fascism and Science

A letter to Signor Rocco, Minister of State, Rome.

My dear Sir,

Two of the most eminent and respected men of science in Italy have applied to me in their difficulties of conscience and requested me to write to you with the object of preventing, if possible, a piece of cruel persecution with which men of learning are threatened in Italy. I refer to a form of oath in which fidelity to the Fascist system is to be promised. The burden of my request is that you should please advise Signor Mussolini to spare the flower of Italy's intellect this humiliation.

However much our political convictions may differ, I know that we agree on one point : in the progressive achievements of the European mind both of us see and love our highest good. Those achievements are based on the freedom of thought and of teaching, on the principle that the desire for truth must take precedence of all other desires. It was this basis alone that enabled our civilization to take its rise in Greece and to celebrate its rebirth in Italy at the Renaissance. This supreme good has been paid for by the martyr's blood of pure and great men, for whose sake Italy is still loved and reverenced to-day.

Far be it from me to argue with you about what inroads on human liberty may be justified by reasons of State. But the pursuit of scientific truth, detached from the practical interests of everyday life, ought to be treated as sacred by every Government, and it is in the highest interests of all that honest servants of truth should be left in peace. This is also undoubtedly in the interests of the Italian State and its prestige in the eyes of the world.

Hoping that my request will not fall on deaf ears, I am, etc. A. E.

Interviewers

To be called to account publicly for everything one has said, even in jest, an excess of high spirits, or momentary anger, fatal as it must be in the end, is yet up to a point reasonable and natural. But to be called to account publicly for what others have said in one's name, when one cannot defend oneself, is indeed a sad predicament. "But who suffers such a dreadful fate?" you will ask. Well, everyone who is of sufficient interest to the public to be pursued by interviewers. You smile incredulously, but I have had plenty of direct experience and will tell you about it.

Imagine the following situation. One morning a reporter comes to you and asks you in a friendly way to tell him something about your friend N. At first you no doubt feel something approaching indignation at such a proposal. But you soon discover that there is no escape. If you refuse to say anything, the man writes: "I asked one of N.'s supposedly best friends about him. But he prudently avoided my questions. This in itself enables the reader to draw the inevitable conclusions." There is, therefore, no escape, and you give the following information: "Mr. N. is a cheerful, straightforward man, much liked by all his friends. He can find a bright side to any situation. His enterprise and industry know no bounds; his job takes up his entire energies. He is devoted to his family and lays everything he possesses at his wife's feet. . . ."

Now for the reporter's version: "Mr. N. takes nothing very seriously and has a gift for making himself liked, particularly as he carefully cultivates a hearty and ingratiating manner. He is so completely a slave to his job that he has no time for the considerations of any non-personal subject or for any mental activity outside it. He spoils his wife unbelievably and is utterly under her thumb. . . ."

A real reporter would make it much more spicy, but I expect this will be enough for you and your friend N. He reads this, and some more like it, in the paper next morning, and his rage against you knows no bounds, however cheerful and benevolent his natural disposition may be. The injury done to him gives you untold pain, especially as you are really fond of him.

What's your next step, my friend? If you know, tell me quickly, so that I may adopt your method with all speed.

Thanks to America

Mr. Mayor, Ladies, and Gentlemen,

The splendid reception which you have accorded to me to-day puts me to the blush in so far as it is meant for me personally, but it gives me all the more pleasure in so far as it is meant for me as a representative of pure science. For this gathering is an outward and visible sign that the world is no longer prone to regard material power and wealth as the highest goods. It is gratifying that men should feel an urge to proclaim this in an official way.

In the wonderful two months which I have been privileged to spend in your midst in this fortunate land, I have had many opportunities of observing what a high value men of action and of practical life attach to the efforts of science; a good few of them have placed a considerable proportion of their fortunes and their energies at the service of scientific enterprises and thereby contributed to the prosperity and prestige of this country.

I cannot let this occasion pass without referring in a spirit of thankfulness to the fact that American patronage of science is not limited by national frontiers. Scientific enterprises all over the civilized world rejoice in the liberal support of American institutions and individuals—a fact which is, I am sure, a source of pride and gratification to all of you.

These tokens of an international way of thinking and feeling are particularly welcome; for the world is to-day more than ever in need of international thinking and feeling by its leading nations and personalities, if it is to progress towards a better and more worthy future. I may be permitted to express the hope that this internationalism of the American nation, which proceeds from a high sense of responsibility, will very soon extend itself to the sphere of politics. For without the active co-operation of the great country of the United States in the business of regulating international relations, all efforts directed towards this important end are bound to remain more or less ineffectual.

I thank you most heartily for this magnificent

reception and, in particular, the men of learning in this country for the cordial and friendly welcome I have received from them. I shall always look back on these two months with pleasure and gratitude.

The University Course at Davos

Senatores boni viri, senatus autem bestia. So a friend of mine, a Swiss professor, once wrote in his irritable way to a university faculty which had annoyed him. Communities tend to be less guided than individuals by conscience and a sense of responsibility. What a fruitful source of suffering to mankind this fact is! It is the cause of wars and every kind of oppression, which fill the earth with pain, sighs, and bitterness.

And yet nothing truly valuable can be achieved except by the unselfish co-operation of many individuals. Hence the man of good will is never happier than when some communal enterprise is afoot and is launched at the cost of heavy sacrifices, with the single object of promoting life and culture.

Such pure joy was mine when I heard about the university courses at Davos. A work of rescue is being carried out there, with intelligence and a wise moderation, which is based on a grave need, though it may not be a need that is immediately obvious to everyone. Many a young man goes to this valley with his hopes fixed on the healing power of its sunny mountains and regains his bodily health. But thus withdrawn for long periods from the will-hardening discipline of normal work and a prey to morbid reflection on his physical condition, he easily loses the power of mental effort and the sense of being able to hold his own in the struggle for existence. He becomes a sort of hot-house plant and, when his body is cured, often finds it difficult to get back to normal life. Interruption of intellectual training in the formative period of youth is very apt to leave a gap which can hardly be filled later.

Yet, as a general rule, intellectual work in moderation, so far from retarding cure, indirectly helps it forward, just as moderate physical work does. It is in this knowledge that the university courses are being instituted, with the object not merely of preparing these young people for a profession but of stimulating them to intellectual activity as such. They are to provide work, training, and hygiene in the sphere of the mind.

Let us not forget that this enterprise is admirably calculated to establish such relations between members of different nations as are favourable to the growth of a common European feeling. The effects of the new institution in this direction are likely to be all the more advantageous from the fact that the circumstances of its birth rule out every sort of political purpose. The best way to serve the cause of internationalism is by co-operating in some life-giving work.

From all these points of view I rejoice that the energy and intelligence of the founders of the university courses at Davos have already attained such a measure of success that the enterprise has outgrown the troubles of infancy. May it prosper, enriching the inner lives of numbers of admirable human beings and rescuing many from the poverty of sanatorium life!

Congratulations to a Critic

To see with one's own eyes, to feel and judge without succumbing to the suggestive power of the fashion of the day, to be able to express what one has seen and felt in a snappy sentence or even in a cunningly wrought word—is that not glorious? Is it not a proper subject for congratulation?

Greeting to G. Bernard Shaw

There are few enough people with sufficient independence to see the weaknesses and follies of their contemporaries and remain themselves untouched by them. And these isolated few usually soon lose their zeal for putting things to rights when they have come face to face with human obduracy. Only to a tiny minority is it given to fascinate their generation by subtle humour and grace and to hold the mirror up to it by the impersonal agency of art. To-day I salute with sincere emotion the supreme master of this method, who has delighted—and educated—us all.

Some Notes on my American Impressions

I must redeem my promise to say something about my impressions of this country. That is not altogether easy for me. For it is not easy to take up the attitude of an impartial observer when one is received with such kindness and undeserved respect as I have been in America. First of all let me say something on this head.

The cult of individual personalities is always, in my view, unjustified. To be sure, nature distributes her gifts variously among her children. But there are plenty of the well-endowed ones too, thank God, and I am firmly convinced that most of them live quiet, unregarded lives. It strikes me as unfair, and even in bad taste, to select a few of them for boundless admiration, attributing superhuman powers of mind and character to them. This has been my fate, and the contrast between the popular estimate of my powers and achievements and the reality is simply grotesque. The consciousness of this extraordinary state of affairs would be unbearable but for one great consoling thought: it is a welcome symptom in an age which is commonly denounced as materialistic, that it makes heroes of men whose ambitions lie wholly in the intellectual and moral sphere. This proves that knowledge and justice are ranked above wealth and power by a large section of the human race. My experience teaches me that this idealistic outlook is particularly prevalent in America, which is usually decried as a particularly materialistic country. After this digression I come to my proper theme, in the hope that no more weight will be attached to my modest remarks than they deserve.

What first strikes the visitor with amazement is the superiority of this country in matters of technics and organization. Objects of everyday use are more solid than in Europe, houses infinitely more convenient in arrangement. Everything is designed to save human labour. Labour is expensive, because the country is sparsely inhabited in comparison with its natural resources. The high price of labour was the stimulus which evoked the marvellous development of technical devices and methods of work. The opposite extreme is illustrated by over-populated China or India, where the low price of labour has stood in the way of the development of machinery. Europe is half-way between the two. Once the machine is sufficiently highly developed it becomes cheaper in the end than the cheapest labour. Let the Fascists in Europe, who desire on narrow-minded political grounds to see their own particular countries more densely populated, take heed of this. The anxious care with which the United States keep out foreign goods by means of prohibitive tariffs certainly contrasts oddly with this notion. . . . But an innocent visitor must not be expected to rack his brains too much, and, when all is said and done, it is not absolutely certain that every question admits of a rational answer.

The second thing that strikes a visitor is the joyous, positive attitude to life. The smile on the faces of the people in photographs is symbolical of one of the American's greatest assets. He is friendly, confident, optimistic, and—without envy. The European finds intercourse with Americans easy and agreeable.

Compared with the American, the European is more critical, more self-conscious, less good-hearted and helpful, more isolated, more fastidious in his amusements and his reading, generally more or less of a pessimist.

Great importance attaches to the material comforts of life, and peace, freedom from care, security are all sacrificed to them. The American lives for ambition, the future, more than the European. Life for him is always becoming, never being. In this respect he is even further removed from the Russian and the Asiatic than the European is. But there is another respect in which he resembles the Asiatic more than the European does: he is less of an individualist than the European—that is, from the psychological, not the economic, point of view.

More emphasis is laid on the "we" than the "I." As a natural corollary of this, custom and convention are very powerful, and there is much more uniformity both in outlook on life and in moral and æsthetic ideas among Americans than among Europeans. This fact is chiefly responsible for America's economic superiority over Europe. Co-operation and the division of labour are carried through more easily and with less friction than in Europe, whether in the factory or the university or in private good works. This social sense may be partly due to the English tradition.

In apparent contradiction to this stands the fact that the activities of the State are comparatively restricted as compared with Europe. The European is surprised to find the telegraph, the telephone, the railways, and the schools predominantly in private hands. The more social attitude of the individual, which I mentioned just now, makes this possible here. Another consequence of this attitude is that the extremely unequal distribution of property leads to no intolerable hardships. The social conscience of the rich man is much more highly developed than in Europe. He considers himself obliged as a matter of course to place a large portion of his wealth, and often of his own energies too, at the disposal of the community, and public opinion, that all-powerful force, imperiously demands it of him. Hence the most important cultural functions can be left to private enterprise, and the part played by the State in this country is, comparatively, a very restricted one.

The prestige of government has undoubtedly been lowered considerably by the Prohibition laws. For nothing is more destructive of respect for the government and the law of the land than passing laws which cannot be enforced. It is an open secret that the dangerous increase of crime in this country is closely connected with this.

There is also another way in which Prohibition, in my opinion, has led to the enfeeblement of the State. The public-house is a place which gives people a chance to exchange views and ideas on public affairs. As far as I can see, people here have no chance of doing this, the result being that the Press, which is mostly controlled by definite interests, has an excessive influence over public opinion.

The over-estimation of money is still greater in this country than in Europe, but appears to me to be on the decrease. It is at last beginning to be realized that great wealth is not necessary for a happy and satisfactory life.

As regards artistic matters, I have been genuinely impressed by the good taste displayed in the modern buildings and in articles of common use; on the other hand, the visual arts and music have little place in the life of the nation as compared with Europe.

I have a warm admiration for the achievements of American institutes of scientific research. We are unjust in attempting to ascribe the increasing superiority of American research-work exclusively to superior wealth; zeal, patience, a spirit of comradeship, and a talent for co-operation play an important part in its successes. One more observation to finish up with. The United States is the most powerful technically advanced country in the world to-day. Its influence on the shaping of international relations is absolutely incalculable. But America is a large country and its people have so far not shown much interest in great international problems, among which the problem of disarmament occupies first place today. This must be changed, if only in the essential interests of the Americans. The last war has shown that there are no longer any barriers between the continents and that the destinies of all countries are closely interwoven. The people of this country must realize that they have a great responsibility in the sphere of international politics. The part of passive spectator is unworthy of this country and is bound in the end to lead to disaster all round.

Reply to the Women of America

An American Women's League felt called upon to protest against Einstein's visit to their country. They received the following answer.

Never yet have I experienced from the fair sex such energetic rejection of all advances; or, if I have, never from so many at once.

But are they not quite right, these watchful citizenesses? Why should one open one's doors to a person who devours hard-boiled capitalists with as much appetite and gusto as the Cretan Minotaur in days gone by devoured luscious Greek maidens, and on top of that is low-down enough to reject every sort of war, except the unavoidable war with one's own wife? Therefore give heed to your clever and patriotic women-folk and remember that the Capitol of mighty Rome was once saved by the cackling of its faithful geese.

II
Politics and Pacifism

Peace

THE importance of securing international peace was recognized by the really great men of former generations. But the technical advances of our times have turned this ethical postulate into a matter of life and death for civilized mankind to-day, and made the taking of an active part in the solution of the problem of peace a moral duty which no conscientious man can shirk.

One has to realize that the powerful industrial groups concerned in the manufacture of arms are doing their best in all countries to prevent the peaceful settlement of international disputes, and that rulers can achieve this great end only if they are sure of the vigorous support of the majority of their peoples. In these days of democratic government the fate of the nations hangs on themselves; each individual must always bear that in mind.

The Pacifist Problem

Ladies and Gentlemen,

I am very glad of this opportunity of saying a few words to you about the problem of pacificism. The course of events in the last few years has once more shown us how little we are justified in leaving the struggle against armaments and against the war spirit to the Governments. On the other hand, the formation of large organizations with a large membership can of itself bring us very little nearer to our goal. In my opinion, the best method in this case is the violent one of conscientious objection, with the aid of organizations for giving moral and material support to the courageous conscientious objectors in each country. In this way we may succeed in making the problem of pacificism an acute one, a real struggle which attracts forceful natures. It is an illegal struggle, but a struggle for people's real rights against their governments in so far as the latter demand criminal acts of the citizen.

Many who think themselves good pacifists will jib at this out-and-out pacifism, on patriotic grounds. Such people are not to be relied on in the hour of crisis, as the World War amply proved.

I am most grateful to you for according me an opportunity to give you my views in person.

Address to the Students' Disarmament Meeting

Preceding generations have presented us, in a highly developed science and mechanical knowledge, with a most valuable gift which carries with it possibilities of making our life free and beautiful such as no previous generation has enjoyed. But this gift also brings with it dangers to our existence as great as any that have ever threatened it.

The destiny of civilized humanity depends more than ever on the moral forces it is capable of generating. Hence the task that confronts our age is certainly no easier than the tasks our immediate predecessors successfully performed.

The foodstuffs and other goods which the world needs can be produced in far fewer hours of work than formerly. But this has made the problem of the division of labour and the distribution of the goods produced far more difficult. We all feel that the free play of economic forces, the unregulated and unrestrained pursuit of wealth and power by the individual, no longer leads automatically to a tolerable solution of these problems. Production, labour, and distribution need to be organized on a definite plan, in order to prevent valuable productive energies from being thrown away and sections of the population from becoming impoverished and relapsing into savagery. If unrestricted *sacro egoismo* leads to disastrous consequences in economic life, it is a still worse guide in international relations. The development of mechanical methods of warfare is such that human life will become intolerable if people do not before long discover a way of preventing war. The importance of this object is only equalled by the inadequacy of the attempts hitherto made to attain it.

People seek to minimize the danger by limitation of armaments and restrictive rules for the conduct of war. But war is not like a parlour-game in which the players loyally stick to the rules. Where life and death are at stake, rules and obligations go by the board. Only the absolute repudiation of all war is of any use here. The creation of an international court of arbitration is not enough. There must be treaties guaranteeing that the decisions of this court shall be made effective by all the nations acting in concert. Without such a guarantee the nations will never have the courage to disarm seriously.

Suppose, for example, that the American, English, German, and French Governments insisted on the Japanese Government's putting an immediate stop to their warlike operations in China, under pain of a complete economic boycott. Do you suppose that any Japanese Government would be found ready to take the responsibility of plunging its country into such a perilous adventure? Then why is it not done? Why must every individual and every nation tremble for their existence? Because each seeks his own wretched momentary advantage and refuses to subordinate it to the welfare and prosperity of the community.

That is why I began by telling you that the fate of the human race was more than ever dependent on its moral strength to-day. The way to a joyful and happy state is through renunciation and self-limitation everywhere.

Where can the strength for such a process come from? Only from those who have had the chance in their early years to fortify their minds and broaden their outlook through study. Thus we of the older generation look to you and hope that you will strive with all your might to achieve what was denied to us.

To Sigmund Freud

Dear Professor Freud,

It is admirable the way the longing to perceive the truth has overcome every other desire in you. You have shown with irresistible clearness how inseparably the combative and destructive instincts are bound up with the amative and vital ones in the human psyche. At the same time a deep yearning for that great consummation, the internal and external liberation of mankind from war, shines out from the ruthless logic of your expositions. This has been the declared aim of all those who have been honoured as moral and spiritual leaders beyond the limits of their own time and country without exception, from Jesus Christ to Goethe and Kant. Is it not significant that such men have been universally accepted as leaders, in spite of the fact that their efforts to mould the course of human affairs were attended with but small success?

I am convinced that the great men—those whose achievements, even though in a restricted sphere, set them above their fellows—are animated to an overwhelming extent by the same ideals. But they have little influence on the course of political events. It almost looks as if this domain, on which the fate of nations depends, had inevitably to be given over to violence and irresponsibility.

Political leaders or governments owe their position partly to force and partly to popular election. They cannot be regarded as representative of the best elements, morally and intellectually, in their respective nations. The intellectual *élite* have no direct influence on the history of nations in these days; their lack of cohesion prevents them from taking a direct part in the solution of contemporary problems. Don't you think that a change might be brought about in this respect by a free association of people whose work and achievements up to date constitute a guarantee of their ability and purity of aim? This international association, whose members would need to keep in touch with each other by a constant interchange of opinions, might, by defining its attitude in the Press—responsibility always resting with the signatories on any given occasion—acquire a considerable and salutary moral influence over the settlement of political questions. Such an association would, of course, be a prey to all the ills which so often lead to degeneration in learned societies, dangers which are inseparably bound up with the imperfection of human nature. But should not an effort in this direction be risked in spite of this? I look upon the attempt as nothing less than an imperative duty.

If an intellectual association of standing, such as I have described, could be formed, it would no doubt have to try to mobilize the religious organizations for the fight against war. It would give countenance to many whose good intentions are paralysed to-day by a melancholy resignation. Finally, I believe that an association formed of persons such as I have described, each highly esteemed in his own line, would be just the thing to give valuable moral support to those elements in the League of Nations which are really working for the great object for which that institution exists.

I had rather put these proposals to you than to anyone else in the world, because you are least of all men the dupe of your desires and because your critical judgment is supported by a most earnest sense of responsibility.

Compulsory Service

From a letter

Instead of permission being given to Germany to introduce compulsory service it ought to be taken away from everybody else: in future none but mercenary armies should be permitted, the size and equipment of which should be discussed at Geneva. This would be better for France than to have to permit compulsory service in Germany. The fatal psychological effect of the military education of the people and the violation of the individual's rights which it involves would thus be avoided.

Moreover, it would be much easier for two countries which had agreed to compulsory arbitration for the settlement of all disputes arising out of their mutual relations to combine their military establishments of mercenaries into a single organization with a mixed staff. This would mean a financial relief and increased security for both of them. Such a process of amalgamation might extend to larger and larger combinations, and finally lead to an "international police," which would be bound gradually to degenerate as international security increased.

Will you discuss this proposal with our friends by way of setting the ball rolling? Of course I do not in the least insist on this particular proposal. But I do think it essential that we should come forward with a positive programme; a merely negative policy is unlikely to produce any practical results.

Germany and France

Mutual trust and co-operation between France and Germany can come about only if the French demand for security against military attack is satisfied. But should France frame demands in accordance with this, such a step would certainly be taken very ill in Germany.

A procedure something like the following seems, however, to be possible. Let the German Government of its own free will propose to the French that they should jointly make representations to the League of Nations that it should suggest to all member States to bind themselves to the following :—

(1) To submit to every decision of the international court of arbitration.

(2) To proceed with all its economic and military force, in concert with the other members of the League, against any State which breaks the peace or resists an international decision made in the interests of world peace.

Arbitration

Systematic disarmament within a short period. This is possible only in combination with the guarantee of all for the security of each separate nation, based on a permanent court of arbitration independent of governments.

Unconditional obligation of all countries not merely to accept the decisions of the court of arbitration but also to give effect to them.

Separate courts of arbitration for Europe with Africa, America, and Asia (Australia to be apportioned to one of these). A joint court of arbitration for questions involving issues that cannot be settled within the limits of any one of these three regions.

The International of Science

At a sitting of the Academy during the War, at the time when national and political infatuation had reached its height, Emil Fischer spoke the following emphatic words : " It's no use, Gentlemen, science is and remains international." The really great scientists have always known this and felt it passionately, even though in times of political confusion they may have remained isolated among their colleagues of inferior calibre. In every camp during the War this mass of voters betrayed their sacred trust. The international society of the academies was broken up. Congresses were and still are held from which colleagues from ex-enemy countries are excluded. Political considerations, advanced with much solemnity, prevent the triumph of purely objective ways of thinking without which our great aims must necessarily be frustrated.

What can right-minded people, people who are proof against the emotional temptations of the moment, do to repair the damage? With the majority of intellectual workers still so excited, truly international congresses on the grand scale cannot yet be held. The psychological obstacles to the restoration of the international associations of scientific workers are still too formidable to be overcome by the minority whose ideas and feelings are of a more comprehensive kind. These last can aid in the great work of restoring the international societies to health by keeping in close touch with like-minded people all over the world and resolutely championing the international cause in their own spheres. Success on a large scale will take time, but it will undoubtedly come. I cannot let this opportunity pass without paying a tribute to the way in which the desire to preserve the confraternity of the intellect has remained alive through all these difficult years in the breasts of a large number of our English colleagues especially.

The disposition of the individual is everywhere better than the official pronouncements. Right-minded people should bear this in mind and not allow themselves to be misled and get angry: *senatores boni viri, senatus autem bestia.*

If I am full of confident hope concerning the progress of international organization in general, that feeling is based not so much on my confidence in the intelligence and high-mindedness of my fellows, but rather on the irresistible pressure of economic developments. And since these depend largely on the work even of reactionary scientists, they too will help to create the international organization against their wills.

The Institute for Intellectual Co-operation

During this year the leading politicians of Europe have for the first time drawn the logical conclusion from the truth that our portion of the globe can only regain its prosperity if the underground struggle between the traditional political units ceases. The political organization of Europe must be strengthened, and a gradual attempt made to abolish tariff barriers. This great end cannot be achieved by treaties alone. People's minds must, above all, be prepared for it. We must try gradually to awaken in them a sense of solidarity which does not, as hitherto, stop at frontiers. It is with this in mind that the League of Nations has created the *Commission de coopération intellectuelle.* This Commission is to be an absolutely international and entirely non-political authority, whose business it is to put the intellectuals of all the nations, who were isolated by the war, into touch with each other. It is a difficult task; for it has, alas, to be admitted that—at least in the countries with which I am most closely acquainted—the artists and men of learning are governed by narrowly nationalist feelings to a far greater extent than the men of affairs.

Hitherto this Commission has met twice a year. To make its efforts more effective, the French Government has decided to create and maintain a permanent Institute for intellectual co-operation, which is just now to be opened. It is a generous act on the part of the French nation and deserves the thanks of all.

It is an easy and grateful task to rejoice and praise and say nothing about the things one regrets or disapproves of. But honesty alone can help our work forward, so I will not shrink from combining criticism with this greeting to the new-born child.

I have daily occasion for observing that the greatest obstacle which the work of our Commission has to encounter is the lack of confidence in its political impartiality. Everything must be done to strengthen that confidence and everything avoided that might harm it.

When, therefore, the French Government sets up and maintains an Institute out of public funds in Paris as a permanent organ of the Commission, with a Frenchman as its Director, the outside observer can hardly avoid the impression that French influence predominates in the Commission. This impression is further strengthened by the fact that so far a Frenchman has also been chairman of the Commission itself. Although the individuals in question are men of the highest reputation, liked and respected everywhere, nevertheless the impression remains.

Dixi et salvavi animam meam. I hope with all my heart that the new Institute, by constant interaction with the Commission, will succeed in promoting their common ends and winning the confidence and recognition of intellectual workers all over the world.

A Farewell

A letter to the German Secretary of the League of Nations

Dear Herr Dufour-Feronce,

Your kind letter must not go unanswered, otherwise you may get a mistaken notion of my attitude. The grounds for my resolve to go to Geneva no more are as follows: Experience has, unhappily, taught me that the Commission, taken as a whole, stands for no serious determination to make real progress with the task of improving international relations. It looks to me far more like an embodiment of the principle *ut aliquid fieri videatur*. The Commission seems to me even worse in this respect than the League taken as a whole.

It is precisely because I desire to work with all my might for the establishment of an international arbitrating and regulative authority *superior to the State*, and because I have this object so very much at heart, that I feel compelled to leave the Commission.

The Commission has given its blessing to the oppression of the cultural minorities in all countries by causing a National Commission to be set up in each of them, which is to form the only channel of communication between the intellectuals of a country and the Commission. It has thereby deliberately abandoned its function of giving moral support to the national minorities in their struggle against cultural oppression.

Further, the attitude of the Commission in the matter of combating the chauvinistic and militaristic tendencies of education in the various countries has been so lukewarm that no serious efforts in this fundamentally important sphere can be hoped for from it.

The Commission has invariably failed to give moral support to those individuals and associations who have thrown themselves without reserve into the business of working for an international order and against the military system.

The Commission has never made any attempt to resist the appointment of members whom it knew to stand for tendencies the very reverse of those it is bound in duty to foster.

I will not worry you with any further arguments, since you will understand my resolve well enough from these few hints. It is not my business to draw up an indictment, but merely to explain my position. If I nourished any hope whatever I should act differently—of that you may be sure.

The Question of Disarmament

The greatest obstacle to the success of the disarmament plan was the fact that people in general left out of account the chief difficulties of the problem. Most objects are gained by gradual steps: for example, the supersession of absolute monarchy by democracy. Here, however, we are concerned with an objective which cannot be reached step by step.

As long as the possibility of war remains, nations will insist on being as perfectly prepared militarily as they can, in order to emerge triumphant from the next war. It will also be impossible to avoid educating the youth in warlike traditions and cultivating narrow national vanity joined to the glorification of the warlike spirit, as long as people have to be prepared for occasions when such a spirit will be needed in the citizens for the purpose of war. To arm is to give one's voice and make one's preparations not for peace but for war. Therefore people will not disarm step by step; they will disarm at one blow or not at all.

The accomplishment of such a far-reaching change in the life of nations presupposes a mighty moral effort, a deliberate departure from deeply ingrained tradition. Anyone who is not prepared to make the fate of his country in case of a dispute depend entirely on the decisions of an international court of arbitration, and to enter into a treaty to this effect without reserve, is not really resolved to avoid war. It is a case of all or nothing.

It is undeniable that previous attempts to ensure peace have failed through aiming at inadequate compromises.

Disarmament and security are only to be had in combination. The one guarantee of security is an undertaking by all nations to give effect to the decisions of the international authority.

We stand, therefore, at the parting of the ways. Whether we find the way of peace or continue along the old road of brute force, so unworthy of our civilization, depends on ourselves. On the one side the freedom of the individual and the security of society beckon to us, on the other slavery for the individual and the annihilation of our civilization threaten us. Our fate will be according to our deserts.

The Disarmament Conference of 1932

I

May I begin with an article of political faith? It runs as follows: The State is made for man, not man for the State. And in this respect science resembles the State. These are old sayings, coined by men for whom human personality was the highest human good. I should shrink from repeating them, were it not that they are for ever threatening to fall into oblivion, particularly in these days of organization and mechanization. I regard it as the chief duty of the State to protect the individual and give him the opportunity to develop into a creative personality.

That is to say, the State should be our servant and not we its slaves. The State transgresses this commandment when it compels us by force to engage in military and war service, the more so since the object and the effect of this slavish service is to kill people belonging to other countries or interfere with their freedom of development. We are only to make such sacrifices to the State as will promote the free development of individual human beings. To any American all this may be a platitude, but not to any European. Hence we may hope that the fight against war will find strong support among Americans.

And now for the Disarmament Conference. Ought one to laugh, weep, or hope when one thinks of it? Imagine a city inhabited by fiery-tempered, dishonest, and quarrelsome citizens. The constant danger to life there is felt as a serious handicap which makes all healthy development impossible. The magistrate desires to remedy this abominable state of affairs, although all his counsellors and the rest of the citizens insist on continuing to carry a dagger in their girdles. After years of preparation the magistrate determines to compromise and raises the question, how long and how sharp the dagger is allowed to be which anyone may carry in his belt when he goes out. As long as the cunning citizens do not suppress knifing by legislation, the courts, and the police, things go on in the old way, of course. A definition of the length and sharpness of the permitted dagger will help only the strongest and most turbulent and leave the weaker at their mercy. You will all understand the meaning of this parable. It is true that we have a League of Nations and a Court of Arbitration. But the League is not much more than a meeting-hall, and the Court has no means of enforcing its decisions. These institutions provide no security for any country in case of an attack on it. If you bear this in mind, you will judge the attitude of the French, their refusal to disarm without security, less harshly than it is usually judged at present.

Unless we can agree to limit the sovereignty of the individual State by all binding ourselves to take joint action against any country which openly or secretly resists a judgment of the Court of Arbitration, we shall never get out of a state of universal anarchy and terror. No sleight of hand can reconcile the unlimited sovereignty of the individual country with security against attack. Will it need new disasters to induce the countries to undertake to enforce every decision of the recognized international court? The progress of events so far scarcely justifies us in hoping for anything better in the near future. But everyone who cares for civilization and justice must exert all his strength to convince his fellows of the necessity for laying all countries under an international obligation of this kind.

It will be urged against this notion, not without a certain justification, that it over-estimates the efficacy of machinery, and neglects the psychological, or rather the moral, factor. Spiritual disarmament, people insist, must precede material disarmament. They say further, and truly, that the greatest obstacle to international order is that monstrously exaggerated spirit of nationalism which also goes by the fair-sounding but misused name of patriotism. During the last century and a half this idol has acquired an uncanny and exceedingly pernicious power everywhere.

To estimate this objection at its proper worth, one must realize that a *reciprocal* relation exists between external machinery and internal states of mind. Not only does the machinery depend on traditional modes of feeling and owe its origin and its survival to them, but the existing machinery in its turn exercises a powerful influence on national modes of feeling.

The present deplorably high development of nationalism everywhere is, in my opinion, intimately connected with the institution of compulsory military service or, to call it by its less offensive name, national armies. A country which demands military service of its inhabitants is compelled to cultivate a nationalistic spirit in them, which provides the psychological foundation of military efficiency. Along with this religion it has to hold up its instrument, brute force, to the admiration of the youth in its schools.

The introduction of compulsory service is therefore, to my mind, the prime cause of the moral collapse of the white race, which seriously threatens not merely the survival of our civilization but our very existence. This curse, along with great social blessings, started with the French Revolution, and before long dragged all the other nations in its train.

Therefore those who desire to encourage the growth of an international spirit and to combat chauvinism must take their stand against compulsory service. Is the severe persecution to which conscientious objectors to military service are subjected to-day a whit less disgraceful to the community than those to which the martyrs of religion were exposed in former centuries? Can you, as the Kellogg Pact does, condemn war and at the same time leave the individual to the tender mercies of the war machine in each country?

If, in view of the Disarmament Conference, we are not to restrict ourselves to the technical problems of organization involved but also to tackle the psychological question more directly from educational motives, we must try on international lines to invent some legal way by which the individual can refuse to serve in the army. Such a regulation would undoubtedly produce a great moral effect.

This is my position in a nutshell: Mere agreements to limit armaments furnish no sort of security. Compulsory arbitration must be supported by an executive force, guaranteed by all the participating countries, which is ready to proceed against the disturber of the peace with economic and military sanctions. Compulsory service, as the bulwark of unhealthy nationalism, must be combated; most important of all, conscientious objectors must be protected on an international basis.

Finally, I would draw your attention to a book, *War again To-morrow*, by Ludwig Bauer, which discusses the issues here involved in an acute and unprejudiced manner and with great psychological insight.

II

The benefits that the inventive genius of man has conferred on us in the last hundred years could make life happy and care-free if organization had been able to keep pace with technical progress. As it is, these hard-won achievements in the hands of our generation are like a razor in the hands of a child of three. The possession of marvellous means of production has brought care and hunger instead of freedom.

The results of technical progress are most baleful where they furnish means for the destruction of human life and the hard-won fruits of toil, as we of the older generation experienced to our horror in the Great War. More dreadful even than the destruction, in my opinion, is the humiliating slavery into which war plunges the individual. Is it not a terrible thing to be forced by the community to do things which every individual regards as abominable crimes? Only a few had the moral greatness to resist; them I regard as the real heroes of the Great War.

There is one ray of hope. I believe that the responsible leaders of the nations do, in the main, honestly desire to abolish war. The resistance to this essential step forward comes from those unfortunate national traditions which are handed on like a hereditary disease from generation to generation through the workings of the educational system. The principal vehicle of this tradition is military training and its glorification, and, equally, that portion of the Press which is controlled by heavy industry and the soldiers. Without disarmament there can be no lasting peace. Conversely, the continuation of military preparations on the present scale will inevitably lead to new catastrophes.

That is why the Disarmament Conference of 1932 will decide the fate of this generation and the next. When one thinks how pitiable, taken as a whole, have been the results of former conferences, it becomes clear that it is the duty of all intelligent and responsible people to exert their full powers to remind public opinion again and again of the importance of the 1932 Conference. Only if the statesmen have behind them the will to peace of a decisive majority in their own countries can they attain their great end, and for the formation of this public opinion each one of us is responsible in every word and deed.

The doom of the Conference would be sealed if the delegates came to it with ready-made instructions, the carrying out of which would soon become a matter of prestige. This seems to be generally realized. For meetings between the statesmen of two nations at a time, which have become very frequent of late, have been used to prepare the ground for the Conference by conversations about the disarmament problem. This seems to me a very happy device, for two men or groups of men can usually discuss things together most reasonably, honestly, and dispassionately when there is no third person present in front of whom they think they must be careful what they say. Only if exhaustive preparations of this kind are made for the Conference, if surprises are thereby ruled out, and an atmosphere of confidence is created by genuine good will, can we hope for a happy issue.

In these great matters success is not a matter of cleverness, still less of cunning, but of honesty and confidence. The moral element cannot be displaced by reason, thank heaven! It is not the individual spectator's duty merely to wait and criticize. He must serve the cause by all means in his power. The fate of the world will be such as the world deserves.

America and the Disarmament Conference

The Americans of to-day are filled with the cares arising out of economic conditions in their own country. The efforts of their responsible leaders are directed primarily to remedying the serious unemployment at home. The sense of being involved in the destiny of the rest of the world, and in particular of the mother country of Europe, is even less strong than in normal times.

But the free play of economic forces will not by itself automatically overcome these difficulties. Regulative measures by the community are needed to bring about a sound distribution of labour and consumption-goods among mankind; without them even the people of the richest country suffocate. The fact is that since the amount of work needed to supply everybody's needs has been reduced through the improvement of technical methods, the free play of economic forces no longer produces a state of affairs in which all the available labour can find employment. Deliberate regulation and organization are becoming necessary to make the results of technical progress beneficial to all.

If the economic situation cannot be cleared up without systematic regulation, how much more necessary is such regulation for dealing with the problems of international politics! Few people still cling to the notion that acts of violence in the shape of wars are either advantageous or worthy of humanity as a method of solving international problems. But they are not logical enough to make vigorous efforts on behalf of the measures which might prevent war, that savage and unworthy relic of the age of barbarism. It requires some power of reflection to see the issue clearly and a certain courage to serve this great cause resolutely and effectively.

Anybody who really wants to abolish war must resolutely declare himself in favour of his own country's resigning a portion of its sovereignty in favour of international institutions: he must be ready to make his own country amenable, in case of a dispute, to the award of an international court. He must in the most uncompromising fashion support disarmament all round, which is actually envisaged in the unfortunate Treaty of Versailles; unless military and aggressively patriotic education is abolished, we can hope for no progress.

No event of the last few years reflects such disgrace on the leading civilized countries of the world as the failure of all disarmament conferences so far; for this failure is due not only to the intrigues of ambitious and unscrupulous politicians, but also to the indifference and slackness of the public in all countries. Unless this is changed we shall destroy all the really valuable achievements of our predecessors.

I believe that the American nation is only

imperfectly aware of the responsibility which rests with it in this matter. People in America no doubt think as follows: "Let Europe go to the dogs, if it is destroyed by the quarrelsomeness and wickedness of its inhabitants. The good seed of our Wilson has produced a mighty poor crop in the stony ground of Europe. We are strong and safe and in no hurry to mix ourselves up in other people's affairs."

Such an attitude is at once base and short-sighted. America is partly to blame for the difficulties of Europe. By ruthlessly pressing her claims she is hastening the economic and therewith the moral collapse of Europe; she has helped to Balkanize Europe, and therefore shares the responsibility for the breakdown of political morality and the growth of that spirit of revenge which feeds on despair. This spirit will not stop short of the gates of America—I had almost said, has not stopped short. Look around, and look forward.

The truth can be briefly stated: The Disarmament Conference comes as a final chance, to you no less than to us, of preserving the best that civilized humanity has produced. And it is on you, as the strongest and comparatively soundest among us, that the eyes and hopes of all are focused.

Active Pacifism

I consider myself lucky in witnessing the great peace demonstration organized by the Flemish people. To all concerned in it I feel impelled to call out in the name of men of good will with a care for the future: "In this hour of opened eyes and awakening conscience we feel ourselves united with you by the deepest ties."

We must not conceal from ourselves that an

improvement in the present depressing situation is impossible without a severe struggle; for the handful of those who are really determined to do something is minute in comparison with the mass of the lukewarm and the misguided. And those who have an interest in keeping the machinery of war going are a very powerful body; they will stop at nothing to make public opinion subservient to their murderous ends.

It looks as if the ruling statesmen of to-day were really trying to secure permanent peace. But the ceaseless piling-up of armaments shows only too clearly that they are unequal to coping with the hostile forces which are preparing for war. In my opinion, deliverance can only come from the peoples themselves. If they wish to avoid the degrading slavery of war-service, they must declare with no uncertain voice for complete disarmament. As long as armies exist, any serious quarrel will lead to war. A pacifism which does not actually try to prevent the nations from arming is and must remain impotent.

May the conscience and the common sense of the peoples be awakened, so that we may reach a new stage in the life of nations, where people will look back on war as an incomprehensible aberration of their forefathers!

Letter to a Friend of Peace

It has come to my ears that in your greatheartedness you are quietly accomplishing a splendid work, impelled by solicitude for humanity and its fate. Small is the number of them that see with their own eyes and feel with their own hearts. But it is their strength that will decide whether the human race must relapse into that hopeless condition which a blind multitude appears to-day to regard as the ideal.

O that the nations might see, before it is too late, how much of their self-determination they have got to sacrifice in order to avoid the struggle of all against all! The power of conscience and the international spirit has proved itself inadequate. At present it is being so weak as to tolerate parleying with the worst enemies of civilization. There is a kind of conciliation which is a crime against humanity, and it passes for political wisdom.

We cannot despair of humanity, since we are ourselves human beings. And it is a comfort that there still exist individuals like yourself, whom one knows to be alive and undismayed.

Another ditto

Dear friend and spiritual brother,

To be quite frank, a declaration like the one before me in a country which submits to conscription in peace-time seems to me valueless. What you must fight for is liberation from universal military service. Verily the French nation has had to pay heavily for the victory of 1918; for that victory has been largely responsible for holding it down in the most degrading of all forms of slavery. Let your efforts in this struggle be unceasing. You have a mighty ally in the German reactionaries and militarists. If France clings to universal military service, it will be impossible in the long run to prevent its introduction into Germany. For the demand of the Germans for equal rights will succeed in the end; and then there will be two German military slaves to every French one, which would certainly not be in the interests of France.

Only if we succeed in abolishing compulsory service altogether will it be possible to educate the youth in the spirit of reconciliation, joy in life, and love towards all living creatures.

I believe that a refusal on conscientious grounds to serve in the army when called up, if carried out by 50,000 men at the same moment, would be irresistible. The individual can accomplish little here, nor can one wish to see the best among us devoted to destruction through the machinery behind which stand the three great powers of stupidity, fear, and greed.

A third ditto

Dear Sir,

The point with which you deal in your letter is one of prime importance. The armament industry is, as you say, one of the greatest dangers that beset mankind. It is the hidden evil power behind the nationalism which is rampant everywhere. . . .

Possibly something might be gained by nationalization. But it is extremely hard to determine exactly what industries should be included. Should the aircraft industry? And how much of the metal industry and the chemical industry?

As regards the munitions industry and the export of war material, the League of Nations has busied itself for years with efforts to get this horrible traffic controlled—with what little success, we all know. Last year I asked a well-known American diplomat why Japan was not forced by a commercial boycott to desist from her policy of force. "Our commercial interests are too strong," was the answer. How can one help people who rest satisfied with a statement like that?

You believe that a word from me would suffice to get something done in this sphere? What an illusion! People flatter me as long as I do not get in their way. But if I direct my efforts towards objects which do not suit them, they immediately turn to abuse and calumny in defence of their interests. And the onlookers mostly keep out of the light, the cowards! Have you ever tested the civil courage of your countrymen? The silently accepted motto is "Leave it alone and don't speak of it." You may be sure that I shall do everything in my power along the lines you indicate, but nothing can be achieved as directly as you think.

Women and War

In my opinion, the patriotic women ought to be sent to the front in the next war instead of the men. It would at least be a novelty in this dreary sphere of infinite confusion, and besides—why should not such heroic feelings on the part of the fair sex find a more picturesque outlet than in attacks on a defenceless civilian?

Thoughts on the World Economic Crisis

If there is one thing that can give a layman in the sphere of economics the courage to express an opinion on the nature of the alarming economic difficulties of the present day, it is the hopeless confusion of opinions among the experts. What I have to say is nothing new and does not pretend to be anything more than the opinion of an independent and honest man who, unburdened by class or national prejudices, desires nothing but the good of humanity and the most harmonious possible scheme of human existence. If in what follows I write as if I were clear about certain things and sure of the truth of what I am saying, this is done merely for the sake of an easier mode of expression; it does not proceed from unwarranted self-confidence or a belief in the infallibility of my somewhat simple intellectual conception of problems which are in reality uncommonly complex.

As I see it, this crisis differs in character from past crises in that it is based on an entirely new set of conditions, due to rapid progress in methods of production. Only a fraction of the available human labour in the world is needed for the production of the total amount of consumption-goods necessary to life. Under a completely free economic system this fact is bound to lead to unemployment. For reasons which I do not propose to analyse here, the majority of people are compelled to work for the minimum wage on which life can be supported. If two factories produce the same sort of goods, other things being equal, that one will be able to produce them more cheaply which employs less workmen—i.e., makes the individual worker work as long and as hard as human nature permits. From this it follows inevitably that, with methods of production what they are to-day, only a portion of the available labour can be used. While unreasonable demands are made on this portion, the remainder is automatically excluded from the process of production. This leads to a fall in sales and profits. Businesses go smash, which further increases unemployment and diminishes confidence in industrial concerns and therewith public participation in these mediating banks; finally the banks become insolvent through the sudden withdrawal of deposits and the wheels of industry therewith come to a complete standstill.

The crisis has also been attributed to other causes which we will now consider.

(1) *Over-production.* We have to distinguish between two things here—real over-production and apparent over-production. By real over-production I mean a production so great that it exceeds the demand. This may perhaps apply to motor-cars and wheat in the United States at the present moment, although even that is doubtful. By " over-production " people usually mean a condition of things in which more of one particular article is produced than can, in existing circumstances, be sold, in spite of a shortage of consumption-goods among consumers. This condition of things I call apparent over-production. In this case it is not the demand that is lacking but the consumers' purchasing-power. Such apparent over-production is only another word for a crisis, and therefore cannot serve as an explanation of the latter; hence people who try to make over-production responsible for the crisis are merely juggling with words.

(2) *Reparations.* The obligation to pay reparations lies heavy on the debtor nations and their industries, compels them to go in for dumping, and so harms the creditor nations too. This is beyond dispute. But the appearance of the crisis in the United States, in spite of the high tariff-wall protecting them, proves that this cannot be the principal cause of the world crisis. The shortage of gold in the debtor countries due to reparations can at most serve as an argument for putting an end to these payments; it cannot

be dragged in as an explanation of the world crisis.

(3) *Erection of new tariff-walls. Increase in the unproductive burden of armaments. Political insecurity owing to latent danger of war.* All these things add considerably to the troubles of Europe, but do not materially affect America. The appearance of the crisis in America shows that they cannot be its principal causes.

(4) *The dropping-out of the two Powers, China and Russia.* This blow to world trade also does not touch America very nearly, and therefore cannot be a principal cause of the crisis.

(5) *The economic rise of the lower classes since the War.* This, supposing it to be a reality, could only produce a scarcity of goods, not an excessive supply.

I will not weary the reader by enumerating further contentions which do not seem to me to get to the heart of the matter. Of one thing I feel certain : this same technical progress which, in itself, might relieve mankind of a great part of the labour necessary to its subsistence, is the main cause of our present troubles. Hence there are those who would in all seriousness forbid the introduction of technical improvements. This is obviously absurd. But how can we find a more rational way out of our dilemma?

If we could somehow manage to prevent the purchasing-power of the masses, measured in terms of goods, from sinking below a certain minimum, stoppages in the industrial cycle such as we are experiencing to-day would be rendered impossible.

The logically simplest but also most daring

method of achieving this is a completely planned economy, in which consumption-goods are produced and distributed by the community. That, in essentials, is what is being attempted in Russia to-day. Much will depend on what results this mighty experiment produces. To hazard a prophecy here would be presumption. Can goods be produced as economically under such a system as under one which leaves more freedom to individual enterprise? Can this system maintain itself at all without the terror that has so far accompanied it, which none of us "westerners" would care to let himself in for? Does not such a rigid, centralized system tend towards protection and hostility to advantageous innovations? We must take care, however, not to allow these suspicions to become prejudices which prevent us from forming an objective judgment.

My personal opinion is that those methods are preferable which respect existing traditions and habits so far as that is in any way compatible with the end in view. Nor do I believe that a sudden transference of the control of industry to the hands of the public would be beneficial from the point of view of production; private enterprise should be left its sphere of activity, in so far as it has not already been eliminated by industry itself in the form of cartelization.

There are, however, two respects in which this economic freedom ought to be limited. In each branch of industry the number of working hours per week ought so to be reduced by law that unemployment is systematically abolished. At the same time minimum wages must be fixed in such a way that the purchasing power of the workers keeps pace with production.

Further, in those industries which have become monopolistic in character through organization on the part of the producers, prices must be controlled by the State in order to keep the creation of new capital within reasonable bounds and prevent the artificial strangling of production and consumption.

In this way it might perhaps be possible to establish a proper balance between production and consumption without too great a limitation of free enterprise, and at the same time to stop the intolerable tyranny of the owners of the means of production (land, machinery) over the wage-earners, in the widest sense of the term.

Culture and Prosperity

If one would estimate the damage done by the great political catastrophe to the development of human civilization, one must remember that culture in its higher forms is a delicate plant which depends on a complicated set of conditions and is wont to flourish only in a few places at any given time. For it to blossom there is needed, first of all, a certain degree of prosperity, which enables a fraction of the population to work at things not directly necessary to the maintenance of life; secondly, a moral tradition of respect for cultural values and achievements, in virtue of which this class is provided with the means of living by the other classes, those who provide the immediate necessities of life.

During the past century Germany has been one of the countries in which both conditions were fulfilled. The prosperity was, taken as a whole, modest but sufficient; the tradition of respect for culture vigorous. On this basis the German nation has brought forth fruits of culture which form an integral part of the development of the modern world. The tradition, in the main, still stands; the prosperity is gone. The industries of the country have been cut off almost completely from the sources of raw materials on which the existence of the industrial part of the population was based. The surplus necessary to support the intellectual worker has suddenly ceased to exist. With it the tradition which depends on it will inevitably collapse also, and a fruitful nursery of culture turn to wilderness.

The human race, in so far as it sets a value on culture, has an interest in preventing such impoverishment. It will give what help it can in the immediate crisis and reawaken that higher community of feeling, now thrust into the background by national egotism, for which human values have a validity independent of politics and frontiers. It will then procure for every nation conditions of work under which it can exist and under which it can bring forth fruits of culture.

Production and Purchasing Power

I do not believe that the remedy for our present difficulties lies in a knowledge of productive capacity and consumption, because this knowledge is likely, in the main, to come too late. Moreover the trouble in Germany seems to me to be not hypertrophy of the machinery of production but deficient purchasing power in a large section of the population, which has been cast out of the productive process through rationalization.

The gold standard has, in my opinion, the serious disadvantage that a shortage in the supply of gold automatically leads to a contraction of credit and also of the amount of currency in circulation, to which contraction prices and wages cannot adjust themselves sufficiently quickly. The natural remedies for our troubles are, in my opinion, as follows :—

(1) A statutory reduction of working hours, graduated for each department of industry, in order to get rid of unemployment, combined with the fixing of minimum wages for the purpose of adjusting the purchasing-power of the masses to the amount of goods available.

(2) Control of the amount of money in circulation and of the volume of credit in such a way as to keep the price-level steady, all special protection being abolished.

(3) Statutory limitation of prices for such articles as have been practically withdrawn from free competition by monopolies or the formation of cartels.

Production and Work

An answer to Cederström

Dear Herr Cederström,

Thank you for sending me your proposals, which interest me very much. Having myself given so much thought to this subject I feel that it is right that I should give you my perfectly frank opinion on them.

The fundamental trouble seems to me to be the almost unlimited freedom of the labour market combined with extraordinary progress in the methods of production. To satisfy the needs of the world to-day nothing like all the available labour is wanted. The result is unemployment and excessive competition among the workers, both of which reduce purchasing power and put the whole economic system intolerably out of gear.

I know Liberal economists maintain that every economy in labour is counterbalanced by an increase in demand. But, to begin with, I don't believe it, and even if it were true, the abovementioned factors would always operate to force the standard of living of a large portion of the human race down to an unnaturally low level.

I also share your conviction that steps absolutely must be taken to make it possible and necessary for the younger people to take part in the productive process. Further, that the older people ought to be excluded from certain sorts of work (which I call " unqualified " work), receiving instead a certain income, as having by that time done enough work of a kind accepted by society as productive.

I too am in favour of abolishing large cities, but not of settling people of a particular type—e.g., old people—in particular towns. Frankly, the idea strikes me as horrible. I am also of opinion that fluctuations in the value of money must be avoided, by substituting for the gold standard a standard based on certain classes of goods selected according to the conditions of consumption—as Keynes, if I am not mistaken, long ago proposed. With the introduction of this system one might consent to a certain amount of " inflation," as compared with the present monetary situation, if one could believe that the State would really make a rational use of the windfall thus accruing to it.

The weaknesses of your plan lie, so it seems to me, in the sphere of psychology, or rather, in your neglect of it. It is no accident that capitalism has brought with it progress not merely in production but also in knowledge. Egoism and competition are, alas, stronger forces than public spirit and sense of duty. In Russia, they say, it is impossible to get a decent piece of bread. . . . Perhaps I am over-pessimistic concerning State and other forms of communal enterprise, but I expect little good from them. Bureaucracy is the death of all sound work. I have seen and experienced too many dreadful warnings, even in comparatively model Switzerland.

I am inclined to the view that the State can only be of real use to industry as a limiting and regulative force. It must see to it that competition among the workers is kept within healthy limits, that all children are given a chance to develop soundly, and that wages are high enough for the goods produced to be consumed. But it can exert a decisive influence through its regulative function if—and there again you are right—its measures are framed in an objective spirit by independent experts.

I would like to write to you at greater length, but cannot find the time.

Minorities

It seems to be a universal fact that minorities —especially when the individuals composing them are distinguished by physical peculiarities—are treated by the majorities among whom they live as an inferior order of beings. The tragedy of such a fate lies not merely in the unfair treatment to which these minorities are automatically subjected in social and economic matters, but also in the fact that under the suggestive influence of the majority most of the victims themselves succumb to the same prejudice and regard their brethren as inferior beings. This second and greater part of the evil can be overcome by closer combination and by deliberate education of the minority, whose spiritual liberation can thus be accomplished.

The efforts of the American negroes in this direction are deserving of all commendation and assistance.

Observations on the Present Situation in Europe

The distinguishing feature of the present political situation of the world, and in particular of Europe, seems to me to be this, that political development has failed, both materially and intellectually, to keep pace with economic necessity, which has changed its character in a comparatively short time. The interests of each country must be subordinated to the interests of the wider community. The struggle for this new orientation of political thought and feeling is a severe one, because it has the tradition of centuries against it. But the survival of Europe depends on its successful issue. It is my firm conviction that once the psychological impediments are overcome the solution of the real problems will not be such a terribly difficult matter. In order to create the right atmosphere, the most essential thing is personal co-operation between men of like mind. May our united efforts succeed in building a bridge of mutual trust between the nations!

The Heirs of the Ages

Previous generations were able to look upon intellectual and cultural progress as simply the inherited fruits of their forebears' labours, which made life easier and more beautiful for them.

But the calamities of our times show us that this was a fatal illusion.

We see now that the greatest efforts are needed if this legacy of humanity's is to prove a blessing and not a curse. For whereas formerly it was enough for a man to have freed himself to some extent from personal egotism to make him a valuable member of society, to-day he must also be required to overcome national and class egotism. Only if he reaches those heights can he contribute towards improving the lot of humanity.

As regards this most important need of the age the inhabitants of a small State are better placed than those of a great Power, since the latter are exposed, both in politics and economics, to the temptation to gain their ends by brute force. The agreement between Holland and Belgium, which is the only bright spot in European affairs during the last few years, encourages one to hope that the small nations will play a leading part in the attempt to liberate the world from the degrading yoke of militarism through the renunciation of the individual country's unlimited right of self-determination.

III
Germany 1933

Manifesto

As long as I have any choice, I will only stay in a country where political liberty, toleration, and equality of all citizens before the law are the rule. Political liberty implies liberty to express one's political views orally and in writing, toleration, respect for any and every individual opinion.

These conditions do not obtain in Germany at the present time. Those who have done most for the cause of international understanding, among them some of the leading artists, are being persecuted there.

Any social organism can become psychically distempered just as any individual can, especially in times of difficulty. Nations usually survive these distempers. I hope that healthy conditions will soon supervene in Germany, and that in future her great men like Kant and Goethe will not merely be commemorated from time to time, but that the principles which they inculcated will also prevail in public life and in the general consciousness.

March, 1933.

Correspondence with the Prussian Academy of Sciences

The following correspondence is here published for the first time in its authentic and complete form. The version published in German newspapers was for the most part incorrect, important sentences being omitted.

The Academy's declaration of April 1, 1933, against Einstein.

The Prussian Academy of Sciences heard with indignation from the newspapers of Albert Einstein's participation in atrocity-mongering in France and America. It immediately demanded an explanation. In the meantime Einstein has announced his withdrawal from the Academy, giving as his reason that he cannot continue to serve the Prussian State under its present Government. Being a Swiss citizen, he also, it seems, intends to resign the Prussian nationality which he acquired in 1913 simply by becoming a full member of the Academy.

The Prussian Academy of Sciences is particularly distressed by Einstein's activities as an agitator in foreign countries, as it and its members have always felt themselves bound by the closest ties to the Prussian State and, while abstaining strictly from all political partisanship, have always stressed and remained faithful to the national idea. It has, therefore, no reason to regret Einstein's withdrawal.

Prof. Dr. Ernst Heymann,
Perpetual Secretary.

Le Coq, near Ostende, *April* 5, 1933

To the Prussian Academy of Sciences,

I have received information from a thoroughly reliable source that the Academy of Sciences has spoken in an official statement of "Einstein's participation in atrocity-mongering in America and France."

I hereby declare that I have never taken any part in atrocity-mongering, and I must add that I have seen nothing of any such mongering anywhere. In general people have contented themselves with reproducing and commenting on the official statements and orders of responsible members of the German Government, together with the programme for the annihilation of the German Jews by economic methods.

The statements I have issued to the Press were concerned with my intention to resign my position in the Academy and renounce my Prussian citizenship; I gave as my reason for these steps that I did not wish to live in a country where the individual does not enjoy equality before the law and freedom to say and teach what he likes.

Further, I described the present state of affairs in Germany as a state of psychic distemper in the masses and also made some remarks about its causes.

In a written document which I allowed the International League for combating Anti-Semitism to make use of for the purpose of enlisting support, and which was not intended for the Press at all, I also called upon all sensible people, who are still faithful to the ideals of a civilization in peril, to do their utmost to prevent this mass-psychosis, which is exhibiting itself in such terrible symptoms in Germany to-day, from spreading further.

It would have been an easy matter for the Academy to get hold of a correct version of my words before issuing the sort of statement about me that it has. The German Press has reproduced a deliberately distorted version of my words, as indeed was only to be expected with the Press muzzled as it is to-day.

I am ready to stand by every word I have published. In return, I expect the Academy to communicate this statement of mine to its members and also to the German public before which I have been slandered, especially as it has itself had a hand in slandering me before that public.

The Academy's Answer of April 11, 1933

The Academy would like to point out that its statement of April 1, 1933, was based not merely on German but principally on foreign, particularly French and Belgian, newspaper reports which Herr Einstein has not contradicted; in addition, it had before it his much-canvassed statement to the League for combating anti-Semitism, in which he deplores Germany's relapse into the barbarism of long-passed ages. Moreover, the Academy has reason to know that Herr Einstein, who according to his own statement has taken no part in atrocity-mongering, has at least done nothing to counteract unjust suspicions and slanders, which, in the opinion of the Academy, it was his duty as one of its senior members to do. Instead of that Herr Einstein has made statements, and in foreign countries at that, such as, coming from a man of world-wide reputation, were bound to be exploited and abused by the enemies not merely of the present German Government but of the whole German people.

For the Prussian Academy of Sciences,

(Signed) H. von Ficker,
E. Heymann,
Perpetual Secretaries.

Berlin, *April* 7, 1933
The Prussian Academy of Sciences.

Professor Albert Einstein, Leyden,
c/o Prof. Ehrenfest, Witte Rosenstr.

Dear Sir,

As the present Principal Secretary of the Prussian Academy I beg to acknowledge the receipt of your communication dated March 28 announcing your resignation of your membership of the Academy. The Academy took cognizance of your resignation in its plenary session of March 30, 1933.

While the Academy profoundly regrets the turn events have taken, this regret is inspired by the thought that a man of the highest scientific authority, whom many years of work among Germans and many years of membership of our society must have made familiar with the German character and German habits of thought, should have chosen this moment to associate himself with a body of people abroad who—partly no doubt through ignorance of actual conditions and events—have done much damage to our German people by disseminating erroneous views and unfounded rumours. We had confidently expected that one who had belonged to our Academy for so long would have ranged himself, irrespective of his own political sympathies, on the side of the defenders of our nation against the flood of lies which has been let loose upon it. In these days of mud-slinging, some of it vile, some of it ridiculous, a good word for the German people from you in particular might have produced a great effect, especially abroad. Instead of which your testimony has served as a handle to the enemies not merely of the present Government but of the German people. This has come as a bitter and grievous disappointment to us, which would no doubt have led inevitably to a parting of the ways even if we had not received your resignation.

Yours faithfully,
(signed) von Ficker.

Le Coq-sur-Mer, Belgium,
April 12, 1933

To the Prussian Academy of Sciences, Berlin.

I have received your communication of the seventh instant and deeply deplore the mental attitude displayed in it.

As regards the fact, I can only reply as follows: What you say about my behaviour is, at bottom, merely another form of the statement you have already published, in which you accuse me of having taken part in atrocity-mongering against the German nation. I have already, in my last letter, characterized this accusation as slanderous.

You have also remarked that a "good word" on my part for "the German people" would have produced a great effect abroad. To this I must reply that such a testimony as you suggest would have been equivalent to a repudiation of all those notions of justice and liberty for which I have all my life stood. Such a testimony would not be, as you put it, a good word for the German nation; on the contrary, it would only have helped the cause of those who are seeking to undermine the ideas and principles which have won for the German nation a place of honour in the civilized world. By giving such a testimony in the present circumstances I should have been contributing, even if only indirectly, to the barbarization of manners and the destruction of all existing cultural values.

It was for this reason that I felt compelled to resign from the Academy, and your letter only shows me how right I was to do so.

Munich, *April* 8, 1933

From the Bavarian Academy of Sciences
to Professor Albert Einstein.

Sir,

In your letter to the Prussian Academy of Sciences you have given the present state of affairs in Germany as the reason for your resignation. The Bavarian Academy of Sciences, which some years ago elected you a corresponding member, is also a German Academy, closely allied to the Prussian and other German Academies; hence your withdrawal from the Prussian Academy of Sciences is bound to affect your relations with our Academy.

We must therefore ask you how you envisage your relations with our Academy after what has passed between yourself and the Prussian Academy.

The President of the Bavarian Academy of Sciences.

Le Coq-sur-Mer, *April* 21, 1933

To the Bavarian Academy of Sciences, Munich.

I have given it as the reason for my resignation from the Prussian Academy that in the present circumstances I have no wish either to be a German citizen or to remain in a position of quasi-dependence on the Prussian Ministry of Education.

These reasons would not, in themselves, involve the severing of my relations with the Bavarian Academy. If I nevertheless desire my name to be removed from the list of members, it is for a different reason.

The primary duty of an Academy is to encourage and protect the scientific life of a country. The learned societies of Germany have, however —to the best of knowledge—stood by and said nothing while a not inconsiderable proportion of German savants and students, and also of professional men of university education, have been deprived of all chance of getting employment or earning their livings in Germany. I would rather not belong to any society which behaves in such a manner, even if it does so under external pressure.

A Reply

The following lines are Einstein's answer to an invitation to associate himself with a French manifesto against Anti-Semitism in Germany.

I have considered this most important proposal, which has a bearing on several things that I have nearly at heart, carefully from every angle. As a result I have come to the conclusion that I cannot take a personal part in this extremely important affair, for two reasons :—

In the first place I am, after all, still a German citizen, and in the second I am a Jew. As regards the first point I must add that I have worked in German institutions and have always been treated with full confidence in Germany. However deeply I may regret the things that are being done there, however strongly I am bound to condemn the terrible mistakes that are being made with the approval of the Government; it is impossible for me to take part personally in an enterprise set on foot by responsible members of a foreign Government. In order that you may appreciate this fully, suppose that a French citizen in a more or less analogous situation had got up a protest against the French Government's action in conjunction with prominent German statesmen. Even if you fully admitted that the protest was amply warranted by the facts, you would still, I expect, regard the behaviour of your fellow-citizen as an act of treachery. If Zola had felt it necessary to leave France at the time of the Dreyfus case, he would still certainly not have associated himself with a protest by German official personages, however much he might have approved of their action. He would have confined himself to—blushing for his countrymen. In the second place, a protest against injustice and violence is incomparably more valuable if it comes entirely from people who have been prompted to it purely by sentiments of humanity and a love of justice. This cannot be said of a man like me, a Jew who regards other Jews as his brothers. For him, an injustice done to the Jews is the same as an injustice done to himself. He must not be the judge in his own case, but wait for the judgment of impartial outsiders.

These are my reasons. But I should like to add that I have always honoured and admired that highly developed sense of justice which is one of the noblest features of the French tradition

IV
The Jews

Jewish Ideals

THE pursuit of knowledge for its own sake, an almost fanatical love of justice, and the desire for personal independence—these are the features of the Jewish tradition which make me thank my stars that I belong to it.

Those who are raging to-day against the ideals of reason and individual liberty and are trying to establish a spiritless State-slavery by brute force rightly see in us their irreconcilable foes. History has given us a difficult row to hoe; but so long as we remain devoted servants of truth, justice, and liberty, we shall continue not merely to survive as the oldest of living peoples, but by creative work to bring forth fruits which contribute to the ennoblement of the human race, as heretofore.

Is there a Jewish Point of View?

In the philosophical sense there is, in my opinion, no specifically Jewish outlook. Judaism seems to me to be concerned almost exclusively with the moral attitude in life and to life. I look upon it as the essence of an attitude to life which is incarnate in the Jewish people rather than the essence of the laws laid down in the Thora and interpreted in the Talmud. To me, the Thora and the Talmud are merely the most important evidence for the manner in which the Jewish conception of life held sway in earlier times.

The essence of that conception seems to me to lie in an affirmative attitude to the life of all creation. The life of the individual has meaning only in so far as it aids in making the life of every living thing nobler and more beautiful. Life is sacred—that is to say, it is the supreme value, to which all other values are subordinate. The hallowing of the supra-individual life brings in its train a reverence for everything spiritual—a particularly characteristic feature of the Jewish tradition.

Judaism is not a creed: the Jewish God is simply a negation of superstition, an imaginary result of its elimination. It is also an attempt to base the moral law on fear, a regrettable and discreditable attempt. Yet it seems to me that the strong moral tradition of the Jewish nation has to a large extent shaken itself free from this fear. It is clear also that "serving God" was equated with "serving the living." The best of the Jewish people, especially the Prophets and Jesus, contended tirelessly for this.

Judaism is thus no transcendental religion; it is concerned with life as we live it and can up to a point grasp it, and nothing else. It seems to me, therefore, doubtful whether it can be called a religion in the accepted sense of the word, particularly as no "faith" but the sanctification of life in a supra-personal sense is demanded of the Jew.

But the Jewish tradition also contains something else, something which finds splendid expression in many of the Psalms—namely, a sort of intoxicated joy and amazement at the beauty and grandeur of this world, of which man can just form a faint notion. It is the feeling from which true scientific research draws its spiritual sustenance, but which also seems to find expression in the song of birds. To tack this on to the idea of God seems mere childish absurdity.

Is what I have described a distinguishing mark of Judaism? Is it to be found anywhere else under another name? In its pure form, nowhere, not even in Judaism, where the pure doctrine is obscured by much worship of the letter. Yet Judaism seems to me one of its purest and most vigorous manifestations. This applies particularly to the fundamental principle of the sanctification of life.

It is characteristic that the animals were expressly included in the command to keep holy the Sabbath day, so strong was the feeling that the ideal demands the solidarity of all living things. The insistence on the solidarity of all human beings finds still stronger expression, and it is no mere chance that the demands of Socialism were for the most part first raised by Jews.

How strongly developed this sense of the sanctity of life is in the Jewish people is admirably illustrated by a little remark which Walter Rathenau once made to me in conversation: "When a Jew says that he's going hunting to amuse himself, he lies." The Jewish sense of the sanctity of life could not be more simply expressed.

Jewish Youth

An Answer to a Questionnaire

It is important that the young should be induced to take an interest in Jewish questions and difficulties, and you deserve gratitude for devoting yourself to this task in your paper. This is of moment not merely for the destiny of the Jews, whose welfare depends on their sticking together and helping each other, but, over and above that, for the cultivation of the international spirit, which is in danger everywhere to-day from a narrow-minded nationalism. Here, since the days of the Prophets, one of the fairest fields of activity has lain open to our nation, scattered as it is over the earth and united only by a common tradition.

Addresses on Reconstruction in Palestine

I

Ten years ago, when I first had the pleasure of addressing you on behalf of the Zionist cause, almost all our hopes were still fixed on the future. To-day we can look back on these ten years with joy; for in that time the united energies of the Jewish people have accomplished a splendid piece of successful constructive work in Palestine, which certainly exceeds anything that we dared to hope then.

We have also successfully stood the severe test to which the events of the last few years have subjected us. Ceaseless work, supported by a noble purpose, is leading slowly but surely to success. The latest pronouncements of the British Government indicate a return to a juster judgment of our case; this we recognize with gratitude.

But we must never forget what this crisis has taught us—namely, that the establishment of satisfactory relations between the Jews and the Arabs is not England's affair but ours. We—that is to say, the Arabs and ourselves—have got to agree on the main outlines of an advantageous partnership which shall satisfy the needs of both nations. A just solution of this problem and one worthy of both nations is an end no less important and no less worthy of our efforts than the promotion of the work of construction itself. Remember that Switzerland represents a higher stage of political development than any national state, precisely because of the greater political problems which had to be solved before a stable community could be built up out of groups of different nationality.

Much remains to be done, but one at least of Herzl's aims has already been realized: its task in Palestine has given the Jewish people an astonishing degree of solidarity and the optimism without which no organism can lead a healthy life.

Anything we may do for the common purpose is done not merely for our brothers in Palestine, but for the well-being and honour of the whole Jewish people.

II

We are assembled to-day for the purpose of calling to mind our age-old community, its destiny, and its problems. It is a community of moral tradition, which has always shown its strength and vitality in times of stress. In all ages it has produced men who embodied the conscience of the Western world, defenders of human dignity and justice.

So long as we ourselves care about this community it will continue to exist to the benefit of mankind, in spite of the fact that it possesses no self-contained organization. A decade or two ago a group of far-sighted men, among whom Herzl of immortal memory stood out above the rest, came to the conclusion that we needed a spiritual centre in order to preserve our sense of solidarity in difficult times. Thus arose the idea of Zionism and the work of settlement in Palestine, the successful realization of which we have been permitted to witness, at least in its highly promising beginnings.

I have had the privilege of seeing, to my great joy and satisfaction, how much this achievement has contributed to the recovery of the Jewish people, which is exposed, as a minority among the nations, not merely to external dangers, but also to internal ones of a psychological nature.

The crisis which the work of construction has had to face in the last few years has lain heavy upon us and is not yet completely surmounted. But the most recent reports show that the world, and especially the British Government, is disposed to recognize the great things which lie behind our struggle for the Zionist ideal. Let us at this moment remember with gratitude our leader Weizmann, whose zeal and circumspection have helped the good cause to success.

The difficulties we have been through have also brought some good in their train. They have shown us once more how strong the bond is which unites the Jews of all countries in a common destiny. The crisis has also purified our attitude to the question of Palestine, purged it of the dross of nationalism. It has been clearly proclaimed that we are not seeking to create a political society, but that our aim is, in accordance with the old tradition of Jewry, a cultural one in the widest sense of the word. That being so, it is for us to solve the problem of living side by side with our brother the Arab in an open, generous, and worthy manner. We have here an opportunity of showing what we have learnt in the thousands of years of our martyrdom. If we choose the right path we shall succeed and give the rest of the world a fine example.

Whatever we do for Palestine we do it for the honour and well-being of the whole Jewish people.

III

I am delighted to have the opportunity of addressing a few words to the youth of this country which is faithful to the common aims of Jewry. Do not be discouraged by the difficulties which confront us in Palestine. Such things serve to test the will to live of our community.

Certain proceedings and pronouncements of the English administration have been justly criticized. We must not, however, leave it at that but learn by experience.

We need to pay great attention to our relations with the Arabs. By cultivating these carefully we shall be able in future to prevent things from becoming so dangerously strained that people can take advantage of them to provoke acts of hostility. This goal is perfectly within our reach, because our work of construction has been, and must continue to be, carried out in such a manner as to serve the real interests of the Arab population also.

In this way we shall be able to avoid getting ourselves quite so often into the position, disagreeable for Jews and Arabs alike, of having to call in the mandatory Power as arbitrator. We shall thereby be following not merely the dictates of Providence but also our traditions, which alone give the Jewish community meaning and stability.

For that community is not, and must never become, a political one; this is the only permanent source whence it can draw new strength and the only ground on which its existence can be justified.

IV

For the last two thousand years the common property of the Jewish people has consisted entirely of its past. Scattered over the wide world, our nation possessed nothing in common except its carefully guarded tradition. Individual Jews no doubt produced great work, but it seemed as if the Jewish people as a whole had not the strength left for great collective achievements.

Now all that is changed. History has set us a great and noble task in the shape of active co-operation in the building up of Palestine. Eminent members of our race are already at work with all their might on the realization of this aim. The opportunity is presented to us of setting up centres of civilization which the whole Jewish people can regard as its work. We nurse the hope of erecting in Palestine a home of our own national culture which shall help to awaken the near East to new economic and spiritual life.

The object which the leaders of Zionism have in view is not a political but a social and cultural one. The community in Palestine must approach the social ideal of our forefathers as it is laid down in the Bible, and at the same time become a seat of modern intellectual life, a spiritual centre for the Jews of the whole world. In accordance with this notion, the establishment of a Jewish university in Jerusalem constitutes one of the most important aims of the Zionist organization.

During the last few months I have been to America in order to help to raise the material basis for this university there. The success of this enterprise was quite natural. Thanks to the untiring energy and splendid self-sacrificing spirit of the Jewish doctors in America, we have succeeded in collecting enough money for the creation of a medical faculty, and the preliminary work is being started at once. After this success I have no doubt that the material basis for the other faculties will soon be forthcoming. The medical faculty is first of all to be developed as a research institute and to concentrate on making the country healthy, a most important item in the work of development. Teaching on a large scale will only become important later on. As a number of highly competent scientific workers have already signified their readiness to take up appointments at the university, the establishment of a medical faculty seems to be placed beyond all doubt. I may add that a special fund for the university, entirely distinct from the general fund for the development of the country, has been opened. For the latter considerable sums have been collected during these months in America, thanks to the indefatigable labours of Professor Weizmann and other Zionist leaders, chiefly through the self-sacrificing spirit of the middle classes. I conclude with a warm appeal to the Jews in Germany to contribute all they can, in spite of the present economic difficulties, for the building up of the Jewish home in Palestine. This is not a matter of charity, but an enterprise which concerns all Jews and the success of which promises to be a source of the highest satisfaction to all.

V

For us Jews Palestine is not just a charitable or colonial enterprise, but a problem of central importance for the Jewish people. Palestine is not primarily a place of refuge for the Jews of Eastern Europe, but the embodiment of the re-awakening corporate spirit of the whole Jewish nation. Is it the right moment for this corporate sense to be awakened and strengthened? This is a question to which I feel compelled, not merely by my spontaneous feelings but on rational grounds, to return an unqualified " yes."

Let us just cast our eyes over the history of the Jews in Germany during the past hundred years. A century ago our forefathers, with few exceptions, lived in the ghetto. They were poor, without political rights, separated from the Gentiles by a barrier of religious traditions, habits of life, and legal restrictions; their intellectual development was restricted to their own literature, and they had remained almost unaffected by the mighty advance of the European intellect which dates from the Renaissance. And yet these obscure, humble people had one great advantage over us: each of them belonged in every fibre of his being to a community in which he was completely absorbed, in which he felt himself a fully privileged member, and which demanded nothing of him that was contrary to his natural habits of thought. Our forefathers in those days were pretty poor specimens intellectually and physically, but socially speaking they enjoyed an enviable spiritual equilibrium.

Then came emancipation, which suddenly opened up undreamed-of possibilities to the individual. Some few rapidly made a position for themselves in the higher walks of business and social life. They greedily lapped up the splendid triumphs which the art and science of the Western world had achieved. They joined in the process with burning enthusiasm, themselves making contributions of lasting value. At the same time they imitated the external forms of Gentile life, departed more and more from their religious and social traditions, and adopted Gentile customs, manners, and habits of thought. It seemed as though they were completely losing their identity in the superior numbers and more highly organized culture of the nations among whom they lived, so that in a few generations there would be no trace of them left. A complete disappearance of Jewish nationality in Central and Western Europe seemed inevitable.

But events turned out otherwise. Nationalities of different race seem to have an instinct which prevents them from fusing. However much the Jews adapted themselves, in language, manners, and to a great extent even in the forms of religion, to the European peoples among whom they lived, the feeling of strangeness between the Jews and their hosts never disappeared. This spontaneous feeling is the ultimate cause of anti-Semitism, which is therefore not to be got rid of by well-meaning propaganda. Nationalities want to pursue their own path, not to blend. A satisfactory state of affairs can be brought about only by mutual toleration and respect.

The first step in that direction is that we Jews should once more become conscious of our existence as a nationality and regain the self-respect that is necessary to a healthy existence. We must learn once more to glory in our ancestors and our history and once again take upon ourselves, as a nation, cultural tasks of a sort calculated to strengthen our sense of the community. It is not enough for us to play a part as individuals in the cultural development of the human race, we must also tackle tasks which only nations as a whole can perform. Only so can the Jews regain social health.

It is from this point of view that I would have you look at the Zionist movement. To-day history has assigned to us the task of taking an active part in the economic and cultural reconstruction of our native land. Enthusiasts, men of brilliant gifts, have cleared the way, and many excellent members of our race are prepared to devote themselves heart and soul to the cause. May every one of them fully realize the importance of this work and contribute, according to his powers, to its success!

The Jewish Community

A speech in London

Ladies and Gentlemen,

It is no easy matter for me to overcome my natural inclination to a life of quiet contemplation. But I could not remain deaf to the appeal of the O.R.T. and O.Z.E. societies[1]; for in responding to it I am responding, as it were, to the appeal of our sorely oppressed Jewish nation.

The position of our scattered Jewish community is a moral barometer for the political world. For what surer index of political morality and respect for justice can there be than the attitude of the

[1] Jewish charitable associations.

nations towards a defenceless minority, whose peculiarity lies in their preservation of an ancient cultural tradition?

This barometer is low at the present moment, as we are painfully aware from the way we are treated. But it is this very lowness that confirms me in the conviction that it is our duty to preserve and consolidate our community. Embedded in the tradition of the Jewish people there is a love of justice and reason which must continue to work for the good of all nations now and in the future. In modern times this tradition has produced Spinoza and Karl Marx.

Those who would preserve the spirit must also look after the body to which it is attached. The O.Z.E. society literally looks after the bodies of our people. In Eastern Europe it is working day and night to help our people there, on whom the economic depression has fallen particularly heavily, to keep body and soul together; while the O.R.T. society is trying to get rid of a severe social and economic handicap under which the Jews have laboured since the Middle Ages. Because we were then excluded from all directly productive occupations, we were forced into the purely commercial ones. The only way of really helping the Jew in Eastern countries is to give him access to new fields of activity, for which he is struggling all over the world. This is the grave problem which the O.R.T. society is successfully tackling.

It is to you English fellow-Jews that we now appeal to help us in this great enterprise which splendid men have set on foot. The last few years, nay, the last few days, have brought us a disappointment which must have touched you in particular nearly. Do not gird at fate, but rather look on these events as a reason for remaining true to the cause of the Jewish commonwealth. I am convinced that in doing that we shall also indirectly be promoting those general human ends which we must always recognize as the highest.

Remember that difficulties and obstacles are a valuable source of health and strength to any society. We should not have survived for thousands of years as a community if our bed had been of roses; of that I am quite sure.

But we have a still fairer consolation. Our friends are not exactly numerous, but among them are men of noble spirit and strong sense of justice, who have devoted their lives to uplifting human society and liberating the individual from degrading oppression.

We are happy and fortunate to have such men from the Gentile world among us to-night; their presence lends an added solemnity to this memorable evening. It gives me great pleasure to see before me Bernard Shaw and H. G. Wells, to whose view of life I am particularly attracted.

You, Mr. Shaw, have succeeded in winning the affection and joyous admiration of the world while pursuing a path that has led many others to a martyr's crown. You have not merely preached moral sermons to your fellows; you have actually mocked at things which many of them held sacred. You have done what only the born artist can do. From your magic box you have produced innumerable little figures which, while resembling human beings, are compact not of flesh and blood, but of brains, wit, and charm. And yet in a way they are more human than we are ourselves, and one almost forgets that they are creations not of Nature, but of Bernard Shaw. You make these charming little figures dance in a miniature world in front of which the Graces stand sentinel and permit no bitterness to enter. He who has looked into this little world sees our actual world in a new light; its puppets insinuate themselves into real people, making them suddenly look quite different. By thus holding the mirror up to us all you have had a liberating effect on us such as hardly any other of our contemporaries has done and have relieved life of something of its earth-bound heaviness. For this we are all devoutly grateful to you, and also to fate, which along with grievous plagues has also given us the physician and liberator of our souls. I personally am also grateful to you for the unforgettable words which you have addressed to my mythical namesake who makes life so difficult for me, although he is really, for all his clumsy, formidable size, quite a harmless fellow.

To you all I say that the existence and destiny of our people depend less on external factors than on ourselves remaining faithful to the moral traditions which have enabled us to survive for thousands of years despite the heavy storms that have broken over our heads. In the service of life sacrifice becomes grace.

Working Palestine

Among Zionist organizations " Working Palestine " is the one whose work is of most direct benefit to the most valuable class of people living there—namely, those who are transforming deserts into flourishing settlements by the labour of their hands. These workers are a selection, made on a voluntary basis, from the whole

Jewish nation, an *élite* composed of strong, confident, and unselfish people. They are not ignorant labourers who sell the labour of their hands to the highest bidder, but educated, intellectually vigorous, free men, from whose peaceful struggle with a neglected soil the whole Jewish nation are the gainers, directly and indirectly. By lightening their heavy lot as far as we can we shall be saving the most valuable sort of human life; for the first settlers' struggle on ground not yet made habitable is a difficult and dangerous business involving a heavy personal sacrifice. How true this is, only they can judge who have seen it with their own eyes. Anyone who helps to improve the equipment of these men is helping on the good work at a crucial point.

It is, moreover, this working class alone that has it in its power to establish healthy relations with the Arabs, which is the most important political task of Zionism. Administrations come and go; but it is human relations that finally turn the scale in the lives of nations. Therefore to support "Working Palestine" is at the same time to promote a humane and worthy policy in Palestine, and to oppose an effective resistance to those undercurrents of narrow nationalism from which the whole political world, and in a less degree the small political world of Palestine affairs, is suffering.

Jewish Recovery

I gladly accede to your paper's request that I should address an appeal to the Jews of Hungary on behalf of Keren Hajessod.

The greatest enemies of the national consciousness and honour of the Jews are fatty degeneration—by which I mean the unconscionableness which comes from wealth and ease—and a kind of inner dependence on the surrounding Gentile world which comes from the loosening of the fabric of Jewish society. The best in man can flourish only when he loses himself in a community. Hence the moral danger of the Jew who has lost touch with his own people and is regarded as a foreigner by the people of his adoption. Only too often a contemptible and joyless egoism has resulted from such circumstances. The weight of outward oppression on the Jewish people is particularly heavy at the moment. But this very bitterness has done us good. A revival of Jewish national life, such as the last generation could never have dreamed of, has begun. Through the operation of a newly awakened sense of solidarity among the Jews, the scheme of colonizing Palestine launched by a handful of devoted and judicious leaders in the face of apparently insuperable difficulties, has already prospered so far that I feel no doubt about its permanent success. The value of this achievement for the Jews everywhere is very great. Palestine will be a centre of culture for all Jews, a refuge for the most grievously oppressed, a field of action for the best among us, a unifying ideal, and a means of attaining inward health for the Jews of the whole world.

Anti-Semitism and Academic Youth

So long as we lived in the ghetto our Jewish nationality involved for us material difficulties and sometimes physical danger, but no social or psychological problems. With emancipation the position changed, particularly for those Jews who turned to the intellectual professions. In school and at the university the young Jew is exposed to the influence of a society with a definite national tinge, which he respects and admires, from which he receives his mental sustenance, to which he feels himself to belong, while it, on the other hand, treats him, as one of an alien race, with a certain contempt and hostility. Driven by the suggestive influence of this psychological superiority rather than by utilitarian considerations, he turns his back on his people and his traditions, and considers himself as belonging entirely to the others while he tries in vain to conceal from himself and them the fact that the relation is not reciprocal. Hence that pathetic creature, the baptized Jewish *Geheimrat* of yesterday and to-day. In most cases it is not pushfulness and lack of character that have made him what he is, but, as I have said, the suggestive power of an environment superior in numbers and influence. He knows, of course, that many admirable sons of the Jewish people have made important contributions to the glory of European civilization; but have they not all, with a few exceptions, done much the same as he?

In this case, as in many mental disorders, the cure lies in a clear knowledge of one's condition and its causes. We must be conscious of our alien race and draw the logical conclusions from it. It is no use trying to convince the others of our spiritual and intellectual equality by arguments addressed to the reason, when their attitude does not originate in their intellects at all. Rather must we emancipate ourselves socially and supply our social needs, in the main, ourselves. We must have our own students' societies and adopt an attitude of courteous but consistent reserve to the Gentiles. And let us live after our own fashion there and not ape duelling and drinking customs which are foreign to our nature. It is possible to be a civilized European and a good citizen and at the same time a faithful Jew who loves his race and honours his fathers. If we remember this and act accordingly, the problem of anti-Semitism, in so far as it is of a social nature, is solved for us.

A Letter to Professor Dr. Hellpach, Minister of State

Dear Herr Hellpach,

I have read your article on Zionism and the Zurich Congress and feel, as a strong devotee of the Zionist idea, that I must answer you, even if it is only shortly.

The Jews are a community bound together by ties of blood and tradition, and not of religion only: the attitude of the rest of the world towards them is sufficient proof of this. When I came to Germany fifteen years ago I discovered for the first time that I was a Jew, and I owe this discovery more to Gentiles than Jews.

The tragedy of the Jews is that they are people of a definite historical type, who lack the support of a community to keep them together. The result is a want of solid foundations in the individual which amounts in its extremer forms to moral instability. I realized that the only possible salvation for the race was that every Jew in the world should become attached to a living society to which the individual rejoiced to belong and which enabled him to bear the hatred and the humiliations that he has to put up with from the rest of the world.

I saw worthy Jews basely caricatured, and the sight made my heart bleed. I saw how schools, comic papers, and innumerable other forces of the Gentile majority undermined the confidence even of the best of my fellow-Jews, and felt that this could not be allowed to continue.

Then I realized that only a common enterprise dear to the hearts of Jews all over the world could restore this people to health. It was a great achievement of Herzl's to have realized and proclaimed at the top of his voice that, the traditional attitude of the Jews being what it was, the establishment of a national home or, more accurately, a centre in Palestine, was a suitable object on which to concentrate our efforts.

All this you call nationalism, and there is something in the accusation. But a communal purpose, without which we can neither live nor die in this hostile world, can always be called by that ugly name. In any case it is a nationalism whose aim is not power but dignity and health. If we did not have to live among intolerant, narrow-minded, and violent people, I should be the first to throw over all nationalism in favour of universal humanity.

The objection that we Jews cannot be proper citizens of the German State, for example, if we want to be a " nation," is based on a misunderstanding of the nature of the State which springs from the intolerance of national majorities. Against that intolerance we shall never be safe, whether we call ourselves a " people " (or " nation ") or not.

I have put all this with brutal frankness for the

sake of brevity, but I know from your writings that you are a man who attends to the sense, not the form.

Letter to an Arab

March 15, 1930

Sir,

Your letter has given me great pleasure. It shows me that there is good will available on your side too for solving the present difficulties in a manner worthy of both our nations. I believe that these difficulties are more psychological than real, and that they can be got over if both sides bring honesty and good will to the task.

What makes the present position so bad is the fact that Jews and Arabs confront each other as opponents before the mandatory power. This state of affairs is unworthy of both nations and can only be altered by our finding a *via media* on which both sides agree.

I will now tell you how I think that the present difficulties might be remedied; at the same time I must add that this is only my personal opinion, which I have discussed with nobody. I am writing this letter in German because I am not capable of writing it in English myself and because I want myself to bear the entire responsibility for it. You will, I am sure, be able to get some Jewish friend of conciliation to translate it.

A Privy Council is to be formed to which the Jews and Arabs shall each send four representatives, who must be independent of all political parties.

Each group to be composed as follows :—

A doctor, elected by the Medical Association;

A lawyer, elected by the lawyers;
A working men's representative, elected by the trade unions;
An ecclesiastic, elected by the ecclesiastics.

These eight people are to meet once a week. They undertake not to espouse the sectional interests of their profession or nation but conscientiously and to the best of their power to aim at the welfare of the whole population of the country. Their deliberations shall be secret and they are strictly forbidden to give any information about them, even in private. When a decision has been reached on any subject in which not less than three members on each side concur, it may be published, but only in the name of the whole Council. If a member dissents he may retire from the Council, but he is not thereby released from the obligation to secrecy. If one of the elective bodies above specified is dissatisfied with a resolution of the Council, it may replace its representative by another.

Even if this "Privy Council" has no definite powers it may nevertheless bring about the gradual composition of differences, and secure as united representation of the common interests of the country before the mandatory power, clear of the dust of ephemeral politics.

Christianity and Judaism

If one purges the Judaism of the Prophets and Christianity as Jesus Christ taught it of all subsequent additions, especially those of the priests, one is left with a teaching which is capable of curing all the social ills of humanity.

It is the duty of every man of good will to strive

steadfastly in his own little world to make this teaching of pure humanity a living force, so far as he can. If he makes an honest attempt in this direction without being crushed and trampled under foot by his contemporaries, he may consider himself and the community to which he belongs lucky.